The Unfortunate Endeavours *of* Charles Henry Brown

The Unfortunate Endeavours *of* Charles Henry Brown:

Aeronaut 1827 – 1870

Terence FitzSimons

PETER LANG
Oxford • Bern • Berlin • Bruxelles • Frankfurt am Main • New York • Wien

Bibliographic information published by Die Deutsche Nationalbibliothek
Die Deutsche Nationalbibliothek lists this publication in the Deutsche Nationalbibliografie; detailed bibliographic data is available on the Internet at http://dnb.d-nb.de.

A catalogue record for this book is available from the British Library.

Library of Congress Cataloging-in-Publication Data

FitzSimons, Terence, 1939-
The unfortunate endeavours of Charles Henry Brown : aeronaut 1827-1870 / Terence FitzSimons.
pages cm
Includes bibliographical references and index.
ISBN 978-3-03-431930-0 (alk. paper)
1. Brown, Charles Henry, 1827-1870. 2. Balloonists--Australia--Biography. 3. Aeronautics--Australia--History--19th century. I. Title.
TL620.B755F58 2015
629.13092--dc23
[B]

2015009924

Cover design by Brian Melville

ISBN: 978-3-0343-1930-0 (print)
ISBN: 978-3-0353-0722-1 (eBook)

Hochfeldstrasse 32, CH-3012 Bern, Switzerland
info@peterlang.com, www.peterlang.com, www.peterlang.net

Contents

CHAPTER 1
Death due to Drowning 1

CHAPTER 2
In the Beginning 15

CHAPTER 3
Hampton 27

CHAPTER 4
Trouble 39

CHAPTER 5
Bad Day at Dublin 49

CHAPTER 6
Bradford Fiasco 57

CHAPTER 7
A New Beginning 73

CHAPTER 8
Royal Gardens 81

CHAPTER 9
Moving On 93

CHAPTER 10
Something in the Air 101

CHAPTER 11
Steady as she Goes 109

CHAPTER 12
The Proposition 115

In London 123

CHAPTER 13
The Overland Route 127

CHAPTER 14
The Simla 135

CHAPTER 15
Eliza's Journey 141

CHAPTER 16
Introduction to Melbourne 147

CHAPTER 17
A Balloon on the Ballaarat 153

CHAPTER 18
Aloft 163

CHAPTER 19
Changing Directions 171

CHAPTER 20
The Dodge 179

CHAPTER 21
The Reprise 185

CHAPTER 22
Hard Times 193

CHAPTER 23
A New Balloon 203

CHAPTER 24
The All England Eleven 211

CHAPTER 25
The Grand Discovery 219

In Melbourne 225

Bibliography 229

Index 235

CHAPTER I

Death due to Drowning

Samuel Curtis Candler was a robust man, and not of an unduly squeamish disposition, but like many he tended to be nonplussed by any incident of sudden and accidental death. So it was that when first appointed as a district coroner he had formed the resolve to adopt a pragmatic approach towards death and corpses. It would be best, he decided, if he were to view a corpse as nothing more than a fleshy shell, an emptied vessel. Any dead body he would consider as mere forensic *residuum*. What it had been in life would be something established by *viva voce* evidence in court. By taking this particular philosophical view he hoped to be well set to deal with his necessary professional contact with dead bodies. His coronial colleagues, Dr Youl and Dr McLean, being medical men, were well used to the odour of death and the gore of lacerated bodies – wounded by accident or the intrusion of their own post-mortem scalpels.[1] But, being honest with himself, Candler had now to admit, after years as district coroner for Prahran, he had failed in his resolve. He was invariably distressed by the sight of the dead, and particularly so in those instances where the deceased had been reduced to a pustulous mass by the inevitable process of decomposition. He much preferred those coronial enquiries he conducted which were investigations into the origins of fires; commercial, nautical or agricultural.[2] No bodies! Unfortunately, however, such investigations did not always necessarily preclude the presence of a corpse. For example, some weeks before Candler had held an inquest into the demise of a certain Mr Downie, who had managed to burn himself to death trying to extinguish a fire in his kerosene store. Good sense would have dictated the abandonment of the burning store and the summoning of the fire brigade along with the acceptance of the loss of some gallons of kerosene. And Downie being a mature man – all of forty-seven years of age – one would have credited him with sufficient common sense to follow such a course of action. He, however, assessed the situation differently, and had been enveloped in flames when a container of fuel exploded as he tried to wrestle it from the blazing shed.[3] Two of the coronial jurors had fainted when viewing the body.

And now, on 6 January 1870 at Tootgarook, a small settlement on the Mornington Peninsula, and situated at an inconvenient distance from his Brunswick residence, Coroner Candler faced further discomfiture. He was due to enquire into the death of Richard Barry, an eighteen-year-old stonemason, whose body had been recovered from the sea at Rye. Candler and his jurors had duly viewed the body, as it lay exposed in a small shed at the rear of the Local Hotel. It was decomposed, with the face, hands and left thigh eaten by fish. Candler, his handkerchief clapped to his nose, had it whispered in his ear by a constable that the deceased would, of course, be identified by his clothing, since there were no features left to assist in this task. Mercifully, none of the jurors had fainted.[4] Chandler ordered the remains to be covered with a tarpaulin, the undertaker was called, and the coroner and his jurors retired to the hotel's smoking room to open the inquest.

On this occasion another factor contributing to Candler's discomfort was the weather. It was extremely hot, and the room of the hostelry in which he now sat with his jury of twelve men, was without benefit of any reasonable ventilation. It was an unpleasant circumstance compounded by the attendance of a clutter of unwashed dusty stonemasons, and the menacing presence of a number of scarified Maori boatmen. Sweltering, Candler listened to the evidence with as much attention as he could muster.

Evidence was given that Barry had been celebrating the festive season with a number of mates, it being Boxing Day. As the evening drew in, Barry left the Tootgarook Hotel intent on returning to his quarters at the Quarantine Station at Point Nepean, where he was engaged on a building project. The route overland to the Station consisted of little more than a rough and ill-defined footpath. Accordingly, the usual way to return to this rugged part of the peninsula was by taking a boat across Port Phillip Bay. The weather at the time had been hot and turbid, with a storm threatening. Already lightning was to be seen over towards Geelong. In these circumstances most of the locals were loath to take their boats out onto the bay but a Maori boatman was finally found who agreed to transport Barry and three of his companions for a fee well above the normal. As anticipated by other Toogarook ferrymen, the storm did break after the group set out. A high wind sprang up and the lowering evening sky was lit by brilliant flashes of lightning. The boat in which Barry and his companions were making the crossing was swamped. The Maori's body was washed up shortly afterward, and Barry's decomposed and bitten remains were found on the beach at Rye eight days later. There was no trace of the other three passengers. But the locals, who knew about such things, reckoned that the tides would eventually wash up whatever remained of the three missing men. For the moment the fate of the others was not the concern of this jury, who dutifully returned a finding of accidental drowning in the inquest of Richard Barry.

As he returned to Brunswick along the uncomfortably rutted and dusty coast road, Coroner Candler silently prayed that he would be spared from further festering corpses over the next little while – or at the very least until the weather turned cooler. His prayer was not answered.

* * *

It was Tuesday 18 January and Jimmy Brown sat in his small dinghy with the float of his fishing line being gently moved by the Yarra river's current. The weather over the past days had been insistently hot, but as he moved his small craft into the shadow of the Prahran railway bridge he was enjoying the cool of the evening. There was no point trying to fish in the morning as gentlemen swimmers, in various stages of undress, had commandeered the river. It was now 9 o'clock and it seemed that all the swimmers had departed. His 'piscatorial endeavours', as Brown liked to refer to his fishing jaunts, had so far yielded no results, but he hoped that in the now undisturbed waters the fish would start to bite. So much for tranquillity. There was a loud splash from the east side of the bridge, and an exasperated Jimmy Brown saw a man's head bobbing about on the surface. His recreation disturbed, Jimmy snapped at the man, 'Are you swimming?' No sooner had he posed the question than it struck him how silly it sounded. What else would the chap be doing in the water?[5] To his surprise the figure waved and called out, 'I say, Sam!' and sank under the surface.

Brown was considered a resourceful man by those who employed him in his trade as a carpenter, and his wit did not desert him now. Some minutes had passed and the erstwhile swimmer had not surfaced. Now quickly reeling in his line, Jimmy rowed to the shore, where he beached his craft and hurried up the south bank hoping for assistance. He looked about for some device to throw as a raft into the water, or maybe a patrolling policeman to whom he could make a report. Glancing back, he could see nothing in the water. As he turned again he spotted a bundle balanced on a wooden rail underneath the bridge. It was in a position where it could not have escaped notice. Brown went quickly to the railing and there discovered that the bundle was a black round-crowned felt hat. The property of the missing swimmer? There was no coat, waistcoat, shirt or boots to be seen, but nestling in the hat were a pipe and tobacco, together with a lead pencil and a quill. By now thoroughly alarmed, Brown quickly relaunched his dinghy. Rowing back across the Yarra he reached the north bank and quickly made his way to the Brunswick police station. There he presented the hat and its contents to the desk sergeant. The sergeant, a large florid faced man, duly noted the report of a suspected drowning in the station day-book, and Brown was dismissed, with the sergeant lugubriously commenting, 'Well, if your chap's drowned, he'll pop up in a while, won't he?' Three days later, at noon, the body of the missing swimmer surfaced in the middle of the Yarra, at a point just opposite the Cremorne Lunatic Asylum.[6]

* * *

Nutt and Murphy, Solicitors, was not a big firm, but they prided themselves on the correctness of their advice to clients and the promptness of their service to contractors, testators and sundry plaintiffs, who sought summons, writs and injunctions.[7] It was therefore annoying in the extreme, the Christmas holiday season now being over and the courts readying themselves once again to sit, that a senior clerk had chosen to absent himself from work. Charles Henry Brown had been with the firm for over five years and his work as a clerk was of the highest standard. He had an excellent hand, and a splendid eye for the appropriate laying out of civil and criminal court forms. Granted, over the past twelve months there had been mornings when Brown had arrived at work in such a state as to arouse a suspicion that he might have over-indulged in liquor the previous night, but the standard of his work was never seen to suffer. But Brown had now been absent for three days and in the circumstances Mr Nutt thought it not inappropriate to inquire as to the circumstances surrounding his clerk's absence from his place of employment. On Friday morning, 21 January, Mr Nutt resolved to call on the Browns. He was well aware of the address and location of Brown's house. There had been, little more than two years ago, all that *fol-der-rol* of time off and late starts while the Brown family had moved from Marion Street to their present address in Carlton Street. At the time Brown had apologised, explained, and drawn little maps illustrating the logistical problems involved in the removal. What Nutt found amusing in the exercise at the beginning had confessedly become tedious in the end. For all of that, Nutt was quite fond of the Yorkshireman.

* * *

Edward Nutt adjusts the venetian blinds and glances out the window of his office as he prepares to set off to the Browns. He is a man of short stature, fair of complexion and stout of build. Fierce sunrays shaft through the slats and it is clear that there has been no abatement of the searing heat of the past days. Nutt has overseen the dispatch of the essential tasks of the day, though again sorely missing the presence of his most adept clerk. It is now five o'clock and he hopes the advance of evening would have dissipated the heat. Clearly, this has not occurred. He makes a decision, for him a daring one, and in the interest of personal comfort, discards his waistcoat!

Carlton Street is an unprepossessing terrace of single storey houses, each barely distinguishable from its neighbour. It is located pleasantly enough at the tree-shaded north end of the Carlton Garden. Because of the heat Nutt takes a cab to his destination. He now pays off the driver and starts down the tree-lined street. While the thoroughfare is in shadow, it is still not cool and Nutt, even *sans* waistcoat, is hot and by now in an ill humour. The heat has coaxed unpleasant odours from the drains and the piles of animal detritus, hosed into the gutters by an enthusiastic street cleaner. He knocks sharply on the door of 193, and is immediately answered by a clatter and a childish shriek. Eliza Brown opens the door even as she snatches at the shirt collar of five-year-old Walter as he tries to make his escape into the street. He is the youngest of the three Brown children still living at home.

'Why, Mister Nutt!' she exclaims, incautiously releasing her grip on young Walter, who triumphantly dashes up the road, galloping in the direction of Murchison Square. Nutt glances after the fleeing child and then turns to Mrs Brown. He notices immediately how drawn she appears.

'I have come to inquire after your husband, Mrs Brown,' he announces politely, removing his hat and placing it across his vestless chest. 'How is he?'

'I beg your pardon, Mr Nutt,' apologises Eliza, stepping back from the door and motioning to Nutt that he should enter. 'Please come in.' The hallway is pleasantly cool. She appears confused and her tone is anxious. 'You are asking after my husband?'

'Please do not upset yourself, Mrs Brown,' says Nutt, still clutching his hat to his chest and now wishing that he were not in such a state of undress. 'I have not come to remonstrate with him concerning his absence, rather to inquire after his health.' Nutt pauses. 'And to ask when may we expect his return.'

'Oh, Mr Nutt, Mr Nutt,' exclaims Eliza, and then to his horror, she starts to weep. Nutt, without his waistcoat, now discovers that he is also without a handkerchief. He is a little flustered.

'Good heavens, Mrs Brown, what is the matter? Is Mr Brown seriously ill? Is he that critically indisposed?' He blurts out the questions, each on the heel of the other. Even in the moment he cannot help but notice that 'critically indisposed' seems a good turn of phrase and he must try to use it in court sometime.

Eliza is herself without a handkerchief and she places the heels of her palms to her eyes and rubs away the tears. She stands silent for a moment and then glances towards the lounge room. 'Charles has been gone since Tuesday past. I have not seen him since 9 o'clock that morning. He was not well then.'

'Oh,' is all Nutt can think to say. He has come prepared to sympathise with a discommoded employee, but he has no desire to be drawn into a matter that he is now imagining to be of an unpleasant and domestic nature. 'When he returns,' says Nutt, moving towards the front door, 'would you please have him come immediately to the office. There is a considerable amount of outstanding documentation which would be most properly attended to by him, rather than a junior clerk.' Nutt is a little disappointed with himself. He realises he has not expressed his concern with sufficient consideration of Mrs Brown's distressed condition. He hesitates at the door for a moment, and, by way of attempting to recover the situation, remarks, 'I hope all will prove to be well.'

As he leaves, Nutt sees young Walter returning from his expedition to the park. It strikes Nutt that the young lad seems singularly undisturbed by his father's unexplained absence.

* * *

On Friday morning, 21 January 1870, Constable Titus Neeson, force number 2178, was on foot patrol. He was making his way west along the south bank of the Yarra River, not that he was suspicious of anything being in any way amiss in that particular area. So far the moment was all serene and Neeson was content to enjoy the quietude, and it was also so much cooler along the riverbank than back in the dusty warrens of Brunswick. The boating brigade would arrive only later in the afternoon, released from the grind of their various work places, and free from the annoying presence of the swimmers.

Titus Neeson, as his name suggests, was Irish. He was a settled family man, with a young son and daughter and another child on the way. He had been accepted into the police at a time when Chief Secretary, Sir John O'Shanassy, was actively encouraging Chief Commissioner Frederick Standish in the recruitment of Irish Catholics into the force.[8] Neeson, content with his lot, was not ambitious and was prepared, where at all possible, to let things tend to themselves. He only interfered in an official capacity when such action was inescapable. As he glanced across the Yarra towards the old Cremorne Gardens, he realised that one of those inescapable occasions had arisen. Floating face down, mid-stream, was a body.

Neeson quickened his pace. The body was being carried slowly down the river towards the solitary punt which plied its trade across the Yarra from a jetty just down from the Sir Henry Barkly Hotel.[9] Later in the day there would be many skiffs but now there was only the punt. Neeson called across to the pole-man, pointing to the body bobbing in the water. The boatman was disinclined to help. Fishing bodies from the Yarra was not an attractive undertaking and besides that he had no wish to miss out on any fare that might chance by. Eventually the boatman agreed to help on the constable's insistence that it was his civic duty. This call on his public conscience was reinforced by the blunt threat that if he did not cooperate Neeson would, from there on in, 'have an eye kept on him'. Thus persuaded, he punted over to collect Neeson. With some effort and the occasional danger of swamping, the pair managed to get the corpse on board the small craft. The next move was simple enough, though unpleasant in its execution. Having landed, Neeson leading with the corpse's feet gripped firm about his waist, and the pole-man grappling with the lolling torso, they made their way straight to the Sir Henry Barkly Hotel. The corpse's face was blue-green and the boatman complained long and loudly of the smell. Their progress attracted the attention of some hardy bystanders, who fell in step and followed their progress.

Mrs Susan Carrick, the hotel's licensee, was sweeping a first floor balcony when she saw the small procession bearing down on her establishment. She became distressed, as she knew what was in store for her.

'Every holder of a publican's license shall at the request of a police officer receive into the house mentioned in such license any dead body that shall be brought to such house for the purpose of an inquest being held thereon,' paraphrased Neeson, as he confronted Mrs Carrick. 'And on conviction for refusing to receive any such body you shall forfeit and pay a sum not exceeding five pounds.'

'I know, I know,' snapped Mrs Carrick. 'Sure, haven't you done this to me before. This isn't good for business you know.' Recovering herself somewhat as she led the laden duo to a back shed, she inquired sourly, 'Where's the quid for keeping the stiff then?'[10]

Neeson stood back and looked with a critical eye at the corpse which they had laid on a horse blanket. He was possessed of a practical disposition. His superior officer considered this a positive attribute – though there were others who felt Neeson was a man who simply lacked feeling. He took out his notebook and pencil. The man lying dead at his feet had clearly been in the water some three or four days and the body was giving off an unpleasant odour. Nonetheless, Neeson wrote 'body fresh' in his notebook, figuring the medical men could deal with such nice questions as degree of decomposition. He was prepared to assess the age of the deceased as being about forty years. The body looked to be about five foot nine or ten inches tall, weighing, Neeson guessed, around twelve stone, with short black hair, turning grey, no whiskers and a little moustache and beard. He was dressed in a black cloth sac coat, black trousers and vest, white shirt, elastic-sided boots and white socks.[11] The note taking done, Neeson pulled the blanket up over the discoloured face, and was about to leave when a thought struck him. He tugged at the body and felt through the coat pockets. Rigor-mortis had long since dissipated, and the arms and head flopped about. In a side pocket was a small square of wood. Neeson took it to the door of the shed the better to see it. It was a neatly engraved medallion with the representation of a balloon, across the gores of which appeared the legend 'Brown'. It looked to be a printer's block. Neeson pocketed it, and shutting the shed door behind him he returned to Richmond Police station.

'What about my quid?' Mrs Carrick called after him.

* * *

The week was not starting well for Samuel Candler. He had been noted to hold an inquiry at the Sir Henry Barkly Hotel on the body of a drowned man, fished from the Yarra some four days previous. The temperature at the Observatory rated as 107.2 degrees Fahrenheit in the sun.[12] Candler was certain he would be faced with a putrefied corpse, and he was also aware that the thorough Dr George Graham had already conducted one of his enthusiastic post mortems. Candler knew he would have to deal once again with the ugliest face of death.

* * *

Transcript: Proceedings of Inquest upon the body of Charles Henry Brown at Richmond 24 January 1870.

Eliza Brown gave evidence on oath:

I have seen the body now shown to the jury. I identify it as that of Charles Henry Brown, my husband, aged 44 years. He was a solicitor's clerk and lived at 193 Carlton Street, Carlton. I last saw him alive on Friday morning last the 18th inst. in his own house, between 8 and 9 o'clock. He was not well – the result I think of too much drink. He was occasionally of intemperate habits. He never threatened to commit self-destruction. He was very low and depressed after drinking. I saw no more of him alive, but I traced him to some public houses up to 6 o'clock that evening. He had no money with him.

(Signed by the witness on the original record of evidence.)

Evidence of James Brown a carpenter residing at Richmond:

On the evening of the 18th instant between 9 and half past 9 o'clock I found this hat just under the iron railway bridge over the Yarra at Richmond. The pipe, pencil, tobacco and quill were in the hat when I found it. The hat was placed apparently so as to attract attention, on the top of a rail. Two or three minutes before I heard a splash in the water about ten yards higher up the river and I saw what I took to be a man in the water. He sang out, 'I say, Sam'. I thought it was a man bathing and called out. The man floated under the bridge and came near where I was fishing. I called out and the body sank. I then walked to the spot where I heard the splash and found the hat. There had been many men bathing there that evening. I went to the police station and reported the occurrence and handed over the hat and the contents.

The evidence of Titus Neeson, Constable stationed at Richmond. Police no. 2178:

On Friday the 21st instant I was on duty on the banks of the Yarra at Richmond and I saw the body of the deceased in the river about the middle of the stream. I got the body out and found the deceased was dressed in respectable clothes. There were no signs of violence about him. On searching the pockets I found only a pocket handkerchief and the piece of wood produced with the name of Brown engraved in a balloon. The body was fresh and did not appear to have been in the water for more than 3 or 4 days.

The evidence of George Graham M.D. a legally qualified medical practitioner residing at Richmond:

I have made a post-mortem examination of the body of Charles Henry Brown. Externally there were no marks of violence perceptible, the body was in an advanced state of decomposition, there was frothy mucus issuing from the mouth and nostrils. The scalp and integuments were very much congested with dark veinous blood. The brain was in a state of pulp from decomposition, the membranes were congested. The lungs were collapsed and very much congested, each plema was adhered to the ribs and contained about a pint of bloody serum. The windpipe was congested on its lining membrane and contained slight traces of frothy membrane. The heart was quite empty, it was fatty. The liver was congested and [illegible] The stomach was quite empty its mucus lining was reddened. The left kidney was diseased, the right one was normal. The bladder was empty. The intestines contained fecal (sic) matter. There was nothing unusual about the other viscera. I believe death was caused by syncopal asphyxia from drowning.[13]

* * *

Candler sat in his office and neatly folded a margin, two fingers wide, into the pages of blue foolscap before him. He entered the appropriate caption and then endorsed the finding of the inquest: 'The Coroner Samuel Curtis Candler and the jury of twelve men returned their finding on oath that – On the 21st January 1870 Charles Henry Brown, in the river Yarra at Richmond, he was found drowned without any marks of violence, but there is not sufficient evidence to show how he became drowned'. Underneath this he carefully scripted in the evidence which he had earlier recorded in a scribbled hand. Mrs Brown had been distressed when giving her evidence and at times had rambled a little in her testimony. He now wrote up her evidence, omitting the questions he had put, and presenting it as a continuous narrative. Tidying up notes as they were transcribed was standard and acceptable practice, though for all his careful application to the task in hand, one or two slips of spelling, punctuation and fact were made in the re-writing.[14] The neatly presented record of proceedings would be sent to the Secretary of the Law Department. The original transcript, with the signed evidence of the witnesses was safely filed away in his office, to be brought out should the law office decide there were grounds for a prosecution. Candler could not imagine how such a thing would occur in this instance, as there was no indication of foul play.

More than a month later the papers from the inquest arrived on the desk of Police Commissioner Frederick Standish. The Law Department passed on all such documents, acting on the premise that the police with their practical and practised eye might spot something criminal in the record that had eluded the less worldly law officers. Standish found nothing untoward in the evidence, and returned the papers with thanks, though he was moved to make the comment: 'This is either a case of accident or suicide. I think the latter the more probable alternative.'

Notes

The only incident in this chapter not directly supported by evidence is Edward Nutt's visit to Eliza Brown. Samuel Curtis Candler, the coroner, lived in Melbourne all his life, and died in 1911 at the ripe old age of 83.

The correspondence of Charles Henry Brown is contained in his notebook and titled by him: *Aeronautica – Correspondence of C.H. Brown With Explanatory Notes etc.* In the endnotes to this work that title is abbreviated to *Correspondence.* The notebook is held in the La Trobe Australian Manuscript Collection , State Library of Victoria, Melbourne. Ref. A629.1332.B.81.

1 *The Argus* January–February 1870, sec. Inquests: various.
2 *Coroners Statute Act 252* 1865 Section 4.
3 *The Argus* 1 January 1870, sec. Inquests: 5.
4 *The Argus* 11 January 1870, sec. Inquests: 5.
5 *The Age* 22 January 1870, sec. News of the Day: 2.
6 VPRS 24/p, *Inquests*, Ref. 78/70 Box 236.
7 Edward Nutt left Australia in 1870. The firm then practiced as Blake and Riggall. Since 1988, following a merger with Dawson Waldron, the company is now known as Blake Dawson Waldron, and has a very large international practice.
8 Robert Haldane, *The People's Force* (Melbourne: Melbourne University Press, 1995) 68.
9 Brian Carroll, *Richmond and East Melbourne Sketchbook* (Melbourne: Rigby, 1976) n.n.
10 *Coroners Statute Act 252* 1865 Section 11.
11 *The Age* 22 January 1870, sec. News of the Day: 2.
12 *The Age* 25 January 1870: 4.

13 VPRS 24/P, *Inquests*, Ref. 78/70 Box 236.

14 Eliza Brown had mis-stated the day on which her husband, Charles Henry Brown, had gone missing. Candler did not pick this up, nor did he clarify the disparity between Neeson's description of the body's state and Graham's comment on the corpse's condition.

CHAPTER 2

In the Beginning

'I am determined I will be an aeronaut, if possible, if I am obliged to make my balloon of paper'. This statement of purpose was set out by Charles Henry Brown in a letter sent in June 1849 from Leeds to the veteran aeronaut, John Hampton. Brown was careful to qualify his statement about taking to the air, and indicate that it was a future objective rather than a present prospect.[1] Hampton at the time was across the Irish Sea in Cork, procuring a living by making balloon ascents from the main cities and towns. He augmented his earnings by presenting the occasional pyrotechnic display along with providing panoramic exhibits relating to topical events. Hampton was flattered by the attention with which young Brown seemed to be following his aeronautical endeavours in Ireland, and the lad was certainly abreast of all the other news from the esoteric world of aerostation.

Leeds in the mid-nineteenth century was a confused place with a multiplicity of architectural styles and a bewildering skyline challenging all sense of order and perspective. Visitors would remark on its visible 'busy-ness' as well as its enthusiastic businesses, and the tremendous movement of all sorts of vehicular traffic on its none-too-wide paved streets.[2] William Osburn, a town overseer and one time trustee of the Leeds Parish Workhouse, commented in verse on the state of the town's waterway, the Aire River:

> The AIRE below is doubly dyed and damned:
> The AIR above, with lurid smoke is crammed,
> The ONE flows steaming foul as Charon's Styx,
> Its poisonous vapours in the other mix.[3]

Whatever the exact nature of the 'poisonous vapours' given off by the Aire, the river would occasionally ignite, to the distress of the locals. While its industrial residue had left it begrimed and grey, and its river noisome, the town was endowed with a lively social and cultural community. With a population of one hundred and seventy thousand, Leeds had a long-established Literary and Philosophical Institution, and a Mechanics' Institute. It strove in all ways to meet the challenging standards set by its neighbouring borough, Bradford, which in the ten years preceding 1849 had come close to doubling its population.[4] Its citizens felt content that they had achieved pre-eminence, for had not the visiting Charles Dickens dubbed Leeds a 'great town'?[5]

It was in this burgeoning town that Benjamin Brown, a hardworking bookkeeper, had married Hannah Higgins at Saint Peter's Anglican Church in April 1823. Brown had attended well to the little schooling he had received. He had disciplined himself to scribe in a good hand, and had learnt as much mathematics as put him ahead of most of the others with whom he had gone to school. Their first child, Edwin, was born on 24 June 1824, and Charles Henry arrived two years later. Benjamin set himself to bring up his sons as best he could – a solid grounding in reading and writing in the Free Grammar School at nearby North Street, and some little skill with numbers. What they did not get to learn at school, he taught them himself at home. Edwin was a solid lad, a plodder, but there was always something a little different about Charles. From the age of five he showed what appeared to his father and mother to be an unhealthy interest in aerostation.[6]

Leeds had had its share of visiting aeronautical adventurers. Vincent Lunardi had passed through the town on his way back from his pioneering flights in Scotland and had taken to the air from the White Cloth Hall on 4 December 1786. Others, such as the veteran aeronaut Charles Green, had visited Leeds to exhibit their aerostatic skills. Just a few months after Benjamin and Hannah were married, William Sadler and Charles Green had staged a balloon race from the town's centre. A large crowd turned out to watch as Sadler took his machine aloft from the Coloured Cloth Hall at the City Square while Green, located a little further up the road at Kirkgate, ascended from the White Cloth Hall. As a pioneer in the use of coal gas, Green opted to have his vehicle filled by the local gas company, while Sadler could only effect a 'mean and miserable' inflation using experimental equipment of his own devising. Green triumphed, taking his craft as far south as Gainsborough.[7] There had been many subsequent balloon ascents, particularly from the Kirkgate during the annual fairs. William Russum, a Leeds aeronaut and balloon builder, was an acquaintance of Benjamin Brown's, and Charles avidly followed Russum's exploits, though he had yet to meet the man.

However it was that young Charles Brown came to catch the aerostatic bug, he was engaged in eager correspondence with some of the craft's leading lights. As early as 1842 he was writing to Mr Davies of Cheltenham requesting details of flights made from the Montpelier Gardens and was also corresponding with Dr Joseph MacSweeny, a Cork medical man, who was an aerostatic enthusiast and had published on the subject.

Brown senior was irritated by his son's enthusiasm. He took a stolid, practical position, which followed the line, 'if there weren't a bob in it, it weren't worth doing', and where was there money to be made in that madcap aeronautical business? Young Charles still continued to hunt after all things connected with matters aerial: newspaper accounts, pamphlets, and such pictures as he could scrounge from old periodicals, as well as the pen and ink drawings of balloons which he had sketched himself.[8] He was also keen to have some of his own material published, and was in correspondence with Henry Coxwell, a London dentist and aeronaut, who was at the time publishing a small aeronautical magazine.[9]

Charles was nearing 19 years of age when Benjamin Brown finally decided to take a concrete step towards putting this whole nonsense of aerostation to rest. He had already forbidden Charles to read any material relating to the subject, not that that had any practical effect, since he was able to gain access to the reading room at the Mechanics' Institute. And forbid it though he did, Benjamin Brown knew that this was a prohibition that his son blithely ignored. Something else had to be done.

* * *

'I've told thee repeatedly, Charlie, this is a fool's game,' stated Benjamin Brown, rapidly tapping the newspaper that lay on his lap with an index finger. 'There's nowt to be gained from folk gallivanting about in balloons.' Brown sat close by the dining room fire. It was early July, but the evenings could still turn nippy.

'But, Dad,' protested Charles, 'aerostation is the way things will go in the future. Canals and railways won't count for that much.'

'Utter rubbish, Charlie, utter rubbish,' snapped his father.

Charles had been over this particular ground with his father time after time, and always the same response. But on each occasion, rather than becoming disenchanted or disillusioned, he found these confrontations only served to fire up his enthusiasm – not that it had ever lagged.

Hannah Brown moved about the kitchen, just out of sight, though not out of earshot, but this was the sort of business best left to her husband and her son to sort out between themselves. Lord only knows they'd had this matter thrashed through time and again, but Charles was not only clever, he was also stubborn, almost to a fault.

'Listen to this,' commanded Mr Brown, lifting up the newspaper and waving it in Charles' direction. 'There is a report here of one of those aeronauts as has come down in the sea off Dublin. What's his name?' He referred to the paper. 'Here we are!' He smoothed out the paper and started to read. 'Mr Gypson started from the Portobello Gardens and came down in the sea off Bray Head. And,' Mr Brown added by way of comment, 'he was rescued by the grace of God and the coast guard.'[10]

'May I see?' asked Charles, stretching out his hand for the newspaper.

'Nay, lad,' snapped his father, 'Thou's never been encouraged by me to read about such nonsense, and I'll not start now.'

Charles was well used to his father's rather theatrical performance with the newspaper, but it was still an irritation. While he was forbidden to read anything on the subject of ballooning, his father would nonetheless gleefully inform him of any accident that came to his notice and read aloud the more grotesque details of any aerial mishap reported in the newspapers.[11]

'Mr Gypson has a parachute, you know,' volunteered Charles. 'He let down a kitten in one once, by way of an experiment.'[12]

'Thou may well smile, Charlie,' said his father, laying aside the newspaper and stretching back in his chair. 'It's little attention thou pays to me, but there's others who can talk against this stupidity.'

Charles shook his head. 'What do you mean, Dad?'

'Thou'll find out soon enough,' replied his father, taking up his newspaper again and starting to read, only to lower it after a moment as he looked over the top of the page at Charles. 'Russum's the man!' he announced cryptically, before returning to his reading.

* * *

So it was that this acquaintance was called upon by Benjamin Brown to perform a favour. William Russum was well known around Leeds and further afield. He had been born in the city in 1806 and had pioneered aerostation there. His first ascent from Leeds was during 1830. The following year he made two spectacular flights, the first marking the Sunday School Jubilee, and then on 8 September 1831, he again took to the air to celebrate the coronation of William IV.[13] For the second of these flights he had specially built the *Coronation Balloon*, a gas vehicle made of silk and cotton. These ascents were made from the White Cloth Hall yard, the site used by William Green for his successful trip aloft in 1823. Again the events attracted large crowds.

But things had not always gone smoothly with Russum's aerial career. He had decided to make a further ascent in the *Coronation Balloon*, this time at Huddersfield just when the local Chartists were holding a rally. This was a rash decision. His action was construed by the rally organisers as an attempt by him, an entrepreneurial capitalist, to distract attention from their cause. The irate Chartists fell upon Russum's balloon and destroyed it. In the chaos he was able to make an escape.[14] Undaunted, he then built a giant of a machine, constructed of varnished muslin gores, and capable of holding 50,000 cubic feet of gas. He named it *Great Northern*, but it too was destroyed, this time in a flying accident.

Young Charles Brown, with his love for statistical information, had collected as much material as he could relating to Russum's flights, and those of his contemporaries: the likes of George and Margaret Graham, Charles Green and George Gale.[15] It was concerning Russum that Charles had written to H. Davis, the proprietor of the Montpelier Gardens, Cheltenham, requesting details of the ascents Russum had made during 1841 in the *Great Northern*. Charles was proud of his personal library of aeronautical cuttings, books and pamphlets, and was very anxious to keep it up to date. This was one of the reasons for his starting to correspond with various aeronauts and the writers and proprietors of aviational publications.

Davis had taken some time in replying to young Brown's query, but when he did write, his letter contained details of another of Russum's mishaps. He reported that Russum's flight in the *Great Northern* had measured up poorly when compared to earlier ascents from the pleasure gardens; he had only managed to stay aloft for fifteen minutes and his balloon had failed to inflate properly due to defects in the envelope – the varnish had rotted the muslin, and to top it all a crowd of no more than five hundred or so had attended the event.[16] This was a bitter disappointment for Davis, since Russum had boasted to him that his flights never failed to attract a crowd of less than 1,500 at Leeds.[17] The news of Russum's misfortune also came as a bitter disappointment to Charles.

This was the man whom Benjamin Brown called on to dissuade Charles from following an aviational avocation. Russum did attempt to do just that, but when the meeting took place and what actually passed between the young man and the older aeronaut is not known. Perhaps it had taken a course not dissimilar to the following:

* * *

(A modest sitting room. A settee and two high wing-backed upholstered chairs down stage center. A half-table with a potted plant atop it stands at the window OP. *(Opposite Prompt will be stage left.) There is a fireplace at the wing* P. *On the mantelpiece there are a number of nick-knacks.* CHARLES BROWN *and his mother,* HANNAH, *are sitting on the settee. They are ill at ease. Up stage there is a net-curtained window. A hint of greenery can be glimpsed through the curtain.* BENJAMIN BROWN *is pacing. There is a knock at the door* OFF *and* OP.*)*

BENJAMIN: That must be Mister Russum now.
(HANNAH *rises and hurries to the door* OP)

BENJAMIN: (*continuing*) There's no need to rush, Mother. This is just a friendly call. It's not the Vicar or owt like that. We don't have to stand on ceremony. Now, lad, thou knows how to behave thyself for our guest, none the less.
(HANNAH *exits* OP)

CHARLES: Yes, Dad, of course I do.

(CHARLES *rises and moves to the fireplace*)

BENJAMIN: Aye, well, it's just as good to mention it. Thou has a way of getting ahead of thyself on occasions, and there's no need to show the family up. Thy brother Edwin has to earn a crust out there as well, and I don't want talk of any uppityness on thy part to get stuck back on him. Thou knows as how people talk.

CHARLES: Oh, Dad. I thought you said there wouldn't be any need to stand on ceremony with Mister Russum. Anyway, what could I do that would cause Edwin any harm?

BENJAMIN: (*heatedly*) There thou go! I told thee; thou's a sharp tongue in thy head, and inclined to use it with more haste than good sense. Mister Russum is coming to help as best he can. So, don't thou set about hindering him before he has a chance to tell thee about this ballooning fad that's caught hold on thee.

(HANNAH *comes from* OP *and ushers in* WILLIAM RUSSUM)

HANNAH: Benjamin, it's Mister Russum come to see you – and Charles.

(WILLIAM RUSSUM *enters. He is in his late forties and of stocky build. He is dressed in a dark suit with a navy cravat.* HANNAH *takes his round crowned hat, and hovers in the background.*)

BENJAMIN: Ah, Mister Russum. 'Tis right good of thee to take time to come and visit us like this. I know thou's a busy man.

RUSSUM: Truth to tell, Mister Brown, I'm not as busy as I'd like to be.

(*looks across at* CHARLES)

So, this is the young man?

CHARLES: Yes, Mister Russum.

RUSSUM: Well, you're a sturdy young fellow. How tall are you, then?

CHARLES: Five foot ten, Sir.

RUSSUM: (*to* BENJAMIN) I'll tell you what, Mister Brown, you've the makings of a fine soldier with this lad of yours.

BENJAMIN: I've no doubt that could be so, but he's got himself a position at the post office, and I reckon pens are better than guns any day of the week.

RUSSUM: You'll get no argument from me on that account.

HANNAH: (*hesitantly*) He has a weakness in the arms, you know.

BENJAMIN: (*sharply*) Put on a pot of tea, Mother.

(HANNAH *exits* OP)

BENJAMIN: (*continuing*) Now, take the weight off thy feet, Mister Russum. I reckon thou's something to say to this son of mine, about this balloon foolery he wants to get himself involved in. No offence, I hope.

RUSSUM (*tartly*) I wouldn't go so far as calling it foolery, Mister Brown.

BENJAMIN: (*nonplussed*) Aye, aye.

RUSSUM: However, point taken, Mister Brown. Risky business, aerostation. Risky business. Sit down, young lad, and we'll have a word.

(*He pauses.* BENJAMIN *moves as if to sit on the settee. A sharp glance from* RUSSUM *stops him*)

BENJAMIN: I'll go and see how that pot of tea is coming along.

(BENJAMIN *exits* OP)

RUSSUM: Now lad, let's get down to cases. You want to be an aeronaut?

CHARLES: Yes sir, more than anything else that I can imagine. I don't mind working in the post office, but that's something I do with an eye to saving money, so that I can buy the necessary material to build my own balloon.

RUSSUM: Come lad, have you any idea what it costs to build a balloon? For that matter, have you any idea how difficult it is to construct an aerostatic craft?

CHARLES: Oh, yes! I've read Doctor MacSweeny's book on the manner of constructing a balloon, and I've studied his system for patterning and cutting gores. I've written to him as well.

RUSSUM: (*impatiently*) Indeed. Well, be mindful, there is a great deal of difference between reading something that has been written in a book and then getting out and getting the thing done, and done properly at that. Books are all well and good, in their place. And I know all that right well, as I've made balloons of my own. But that aside, it's a dangerous and sometimes mucky business – aerostation.

CHARLES: I know you've built balloons – the *Coronation* and the *Great Northern*. And I'm aware of the dangers involved in flying.

RUSSUM: Are you, indeed? I don't mean such things as a balloon ripping when you are at twenty thousand feet. Do you know, I've been physically attacked, not once, but a number of times.

CHARLES: Like when you were attacked by the Chartists at Huddersfield?

RUSSUM: You know about that then?

CHARLES: Yes. They destroyed the *Coronation*.

RUSSUM: And gave me a right thumping to boot. It was only Mister O'Connor, from the *Northern Star* newspaper, who knew what it was all really about, and that I was tending to my business with no evil intent towards the Chartists. He kept the mob back long enough for me to get out of it in the end.

CHARLES: Mister O'Connor wrote about it in his paper later. I read it down at the Mechanics' Institute.

RUSSUM: Oh, so you attend at the Mechanics' Institute, do you?

CHARLES: Yes, and they have copies of some of the aerostatic magazines. I read Mister Well's *Balloon* there, and I've written to him.

RUSSUM: Indeed, you have!

(RUSSUM *rises and moves up stage. He draws aside the curtain slightly and looks out of the window. There is silence for a moment. He drops the net curtain and turns.*)

RUSSUM: (*continuing*) This part of the world has a fair reputation in matters aeronautic. Mister Lunardi flew from down at the White Linen Hall when he was on his way back from Scotland, and that was long before I was born, but there has never been any money in it.

CHARLES: Surely, if it were done properly.

RUSSUM: (*testily*) I've been trying long enough, young lad, and hard enough. And I know of none that's made over much money out of this craft, save for old Charlie Green. Then he never put his hand in his pocket for anything but that he had to.

CHARLES: Mister Davies at Montpelier Gardens, told me you'd not a good ascent the last time you tried from there.

RUSSUM: Davies told you?

CHARLES: I wrote to him, when I heard you were to visit. I inquired as to your ascents from his gardens.

RUSSUM: Did you indeed? And what in the name of heaven would that man know about a good ascent or otherwise? And Montpelier Gardens – pshaw! That's the place where they had cripples playing cricket, and a riot to boot.

CHARLES: Surely that was many years ago, and Mister Davies had nothing to do with that.

RUSSUM: Mister Davies is out to make as much money out of the likes of me as he can manage. That man's as close to being a huckster as makes no difference. The point is, and you will have to take my word on this, for it would be a terrible shame if you were to go so far as to seek to prove otherwise: there is no money to be had in ballooning. You're blessed with a steady situation at the post office. You can be set up for life there, so don't throw such a fine position away. I wish I had the opportunity to be permanent and pensionable.

CHARLES: Mister Russum, I don't want to be permanent and pensionable. I want to take to the air.

RUSSUM: Have you any idea what it costs to fill a balloon with gas?

CHARLES: No. But I believe Mister Graham and Mister Green and others have obtained special discounted rates from gas companies. But if such costs were prohibitive I would build a Montgolfier.

RUSSUM: A fire-balloon! And burn yourself to death?

CHARLES: But, Mister Russum, you've taken up a fire balloon, many times.

RUSSUM: Aye, and not ever again, you can trust me on that.

CHARLES: I honestly believe, if properly managed, a balloon ascent can prove to be a financially rewarding undertaking.

RUSSUM: (*sharply*) If properly managed?

(*He is distracted by noise off at* OP. *Crockery rattles and a door is opened and closed.* HANNAH *enters with a tea tray.* BENJAMIN *follows. There is enough space on the side table to set the tray and serve the tea.*)

BENJAMIN: There we are. It seems thee's had thyselves a nice little chat.

RUSSUM: Well, Mister Brown, and Missus Brown, you've got yourselves a strong-willed lad here.

BENJAMIN: Aye. But he's seen sense, hasn't he?

RUSSUM: I honestly wouldn't know, Mister Brown.

(HANNAH *goes to* RUSSUM *at* OP *and hands him a cup of tea.* BENJAMIN *stands stage centre and glowers at* CHARLES. HANNAH *moves across to* BENJAMIN.)

HANNAH: (*sotto voce*) Leave the lad alone, Benjamin. Leave the lad alone.[18]

* * *

Charles felt he had justified himself, if not to his father, then he considered he had at least established a valid case with William Russum. Russum for his part felt that he had been dealing with an impertinent young upstart, who would in time receive his just deserts, if he persisted in his ill-considered purpose of becoming an aeronaut. Persist Charles did.

Notes

Brown's older brother, Edwin, does not appear on the scene during this period. It is only when Brown has moved to Australia and the two brothers start to correspond that he begins to move into the story. Brown's parents move off the stage entirely.

* * *

1 Brown to Hampton, 16 July 1849, *Correspondence*: 20.
2 Gordon Stables, *Leaves from the log of a Gentleman Gipsy* (London: Jarrod & Sons, 1891) 99.
3 Asa Briggs, *Victorian Cities* (Harmondsworth: Penguin Books, 1980) 139.
4 'These Crowded Hives' Population figures. 24 May 2001. <www.history.unimelbourne.edu.au/industrial britain>.
5 'Speeches: Literary and Social by Charles Dickens' *Leeds, 1 December 1849*. 29 May 2001. <http://www.underthes/Dickens/speeches/speeches45.htm>.
6 Brown to Hampton, 18 June 1849, *Correspondence*: 16.
7 *Leodis-Sights,* 'Read All About It', 13 May 2001. <*http://*www..dnausers.d-n-a.net/>.
8 Brown to Delcourt, n.d. December 1850, *Correspondence*: 123.
9 Coxwell to Brown, 14 August 1845, *Correspondence*: 2.
10 F.E. Dixon, 'Ballooning in Dublin', *Dublin Historical Record.Vol.XIV*. (Dublin: Old Dublin Society. 1955): 10.
11 Brown to MacSweeny, 24 October 1863, *Correspondence*: 405.
12 Marshall to Brown, 26 July 1841, *Correspondence*: 36.
13 The Summergill Collection. Vol.3: 34. Q942.75 (Leeds Reference Library)
14 Brown to Hampton, 18 June 1849, *Correspondence*: 14.
15 Lionel Rolt, *The Aeronauts* (London: Longman, 1966) passim.
16 Davis to Brown, 4 April 1843, *Correspondence*: 2.
17 Brown to Hampton, 18 June 1849, *Correspondence*: 14.
18 The meeting with Russum took place at an unspecified date in Leeds. Brown recalled the event, and his father's conduct apropos newspaper reports on aerostation, when writing to Dr MacSweeny many years later, *Correspondence*, 24 October 1863: 405. The strange cricket match at Montpelier Gardens – 'cripples playing cricket' – is recorded in Hibbert's *Social History in Victorian Britain* (Melbourne: Angus & Robertson, 1975) 99. It took place during 1796, long before Mr Davies was associated with the establishment. One team was made up of one-armed players, while the opposition consisted of players who were minus a leg.

CHAPTER 3

Hampton

Charles Brown's first letter to John Hampton was written in June 1849, at least two years after William Russum's failed attempt to convince the young man that he should abandon his plan of become an aeronaut. The initial contact had started with a letter from Brown inquiring of Hampton if he had written any works on aerostation. He hadn't; and he was the first to admit that he was not an adept with the pen. Brown had determined to adopt Hampton as his mentor, though he did not then, or subsequently, declare that to be his purpose. Later he would cast Henry Coxwell, a London aviator, in that role, but for the moment Hampton held his full attention.

Hampton was twenty-seven years older than Brown, and had served in the Royal Navy before turning to ballooning as a profession.[1] He was greatly experienced in the field of aerostation, being also a pioneer parachutist.[2] Hampton was quick enough to reply to Charles' letter. In his response he revealed that he had contemplated going to Leeds to undertake an ascent, but was unsure whether the venture would be a paying proposition. He suggested to Charles that he might like to make some enquiries in this regard. Brown, in his reply, assured Hampton that Leeds and Bradford would be paying propositions; Leeds because balloon ascents had never failed to draw less than a crowd of 1,500 persons to the cloth Hall Yard; and Bradford, since it was a rapidly developing town from which only one balloon ascent had ever been made, as far back as 1828. Brown took the opportunity of boasting to Hampton that he had invented a method of fireproofing Montgolfier fire-balloons, and of his design for a new style of parachute. Hampton may or may not have been interested in these revelations, but he let them pass without comment.[3]

Hampton had made a goodly number of his ascents from Irish soil. He had been ballooning there since 1844 and had more or less a monopoly in the field of aerostation. 'I have been treated very kind (sic) by the Irish people – to know them you must live among them', he wrote to Brown.[4] This was a charitable rather than an accurate statement, since on a number of occasions the locals had set about his balloons when they landed, and in at least one instance had done considerable damage to the vehicle.[5] But then it was a fact that many an aeronaut in England and on the continent had fallen victim to the physical assault of mobs and peasantry; the former were usually upset by the failure of a balloon to get off the ground, while the latter were, in the main, incensed by the manner in which a balloon had come to ground, landing amongst their crops or flocks.

To add spice to his ascents, Hampton had devised a system of 'rewards' calculated to ensure his safe return to dry land should his vehicle come down in the Irish Sea. He offered cash payments of 25/- to the first boatman to reach him, and then 15/-, 10/- and 5/- respectively for the next three on the scene.[6] At times this offer was interpreted by some of the populace as extending to the 'rescue' of Hampton's balloon when it came down on dry land. This may in part explain why his balloon, *Erin Go Bragh*, occasionally came under attack upon landing!

Hampton had started his ballooning career during 1838. At one time his only rival for pre-eminence in the air was the veteran Charles Green. His first balloon, *Albion*, was lost in a fire. He quickly built a new vehicle and, with an eye to his Irish audience, cannily named it *Erin Go Bragh*, roughly translated as 'Ireland forever!' It was second only in size in the United Kingdom to Green's famous *Nassau*.[7]

Charles Brown's opening correspondence with Hampton had been directed to him care of the Rotunda Gardens, Dublin. The city boasted a number of centres of open-air amusement at the time, with the Rotunda and the Portobello Gardens being the premier establishments for balloon ascents. Dublin, however, was the site of some other venues that in the past had hosted aeronauts. Belvedere House, former residence of the Anglican Primate, Richard Robinson, was one such place. It and Clontarf House were at times thrown open to the public and various forms of entertainment presented. William Sadler had taken off from Belvedere House on 1 October 1812, and within 35 minutes was off the coast of Wales, but unfortunately his balloon came down in the sea.[8] His son would many years later undertake a similar trip with the same result, and in the process would provide Benjamin Brown with one more news item of aeronautical misadventure with which to taunt his son.

Hampton preferred to fly from the more purpose-built locations of the Rotunda and Portobello Gardens. Portobello Gardens, situated in the fashionable suburb of Rathmines, had opened in 1839, adjacent to the Grand Canal that carried cargo and passengers between Dublin and the Shannon Harbour. The Grand Canal Hotel near the Gardens provided a regular clientele from amongst those people waiting to take passage on the canal barges.[9] The Rotunda Gardens and Assembly Rooms were located at the north end of Sackville Street, Dublin's central boulevard. Designed in imitation of the Vauxhall and Ranelagh Gardens in London, the revenue from the Gardens was used to provide financial support for the Lying-in-Hospital, which shared the same site.[10]

The use of specially prepared sites did not, of course, furnish a guarantee that all would go well with a flight. A much puffed attempt by Hampton to fly across the Irish Channel was scheduled to take place on 21 June 1848, from the Rotunda Gardens, and an invitation was extended to 'any gentleman of light weight' who wished to accompany the aeronaut on this adventure, to inquire as to terms at the office.[11] On the due date, there was a large crowd in attendance inside the Gardens. Many more looked on from the windows of the adjacent hospital. Accompanied by Sandham Symes Esq. and Mr Bateman, vocalist at the *Shades Tavern*, Hampton took the balloon clear of Dublin but failed to get out to sea. The group came down at a property ten miles north of the city. During the process of deflating the balloon, it got loose and rose to a great height. Though chased vigorously on horseback, it was not retrieved at the time; the wreckage was found some days later at Dunleek, Co. Meath.[12] This escapee was in fact the *Erin Go Bragh*, but within three weeks Hampton built a replacement using the original plans, and the balloon was promptly given the same name as its predecessor.[13]

Meanwhile Brown had been making inquiries in Leeds as to a possible site from which Hampton could make an ascent. Both of the Cloth Halls had proved successful venues in the past. Hampton was very happy to have an unpaid agent acting for him on the other side of the Irish Sea – a young man who obviously knew a thing or two about aerostation. He wrote to Brown, this time from Cork, and thanked him for his offer to find a venue at Leeds or thereabout, suitable for a profitable ascent. At the same time he sent leaflets relating to his proposed flight from the Cork Corn Exchange. 'You will see that I ascend in Cork on the 16th [July] and please God the weather is fine I shall I think have a bumper'.[14] In response to this letter Brown excitedly replied with the news that he had in hand the writing of a work on aerostation and in addition he had completed the manufacture of the top portion of a Montgolfier balloon, though work had presently halted for want of funds. 'I am very poor, and have a very small income', he wrote pathetically.[15] If this was intended to stir within Hampton's breast some charitable or philanthropic response, it failed. Hampton was not so rash as to encourage competition in his chosen field, even if it were as yet manifest only in the gushing of an impecunious amateur. Moreover, Hampton at this time had serious problems of his own.

The *Erin Go Bragh* was to go aloft on Monday, 16 July 1849. The event was under the patronage of the Mayor of Cork, along with Colonel Pole of the 12th Lancers, Colonel Hemphill of the 26th Cameronians, and J. Roche Esq., President of the Cork Royal Institution. Hampton had certainly caught the attention of the local leading lights, though curiously the noted local enthusiast and author of aeronautical works, Dr Joseph MacSweeny, was not named as being involved with the ascent.[16] On the appointed day, with two military bands playing lustily, Hampton prepared for the ascent, but with a capacity of 30,000 cubic feet, *Erin Go Bragh* proved too great a challenge for the gas company to meet at that time, and the attempt to get the vessel off the ground failed. It was quickly arranged for a sufficient supply of gas to be provided the following day, and the crowd departed in relatively good humour. In anticipation of the next attempt, some wit posted placards about the town announcing that Hampton had been engaged by the Pope to fly! Tuesday was a less clement day, with an occasional high gust of wind. This time the gas company met its undertaking and with the envelope fully inflated Hampton prepared to go aloft. The pre-flight process involved the gradual lowering of weights attached to the balloon's netting. These weights were brought down mesh by mesh and discarded, thus increasing the buoyancy of the craft. All was proceeding well until a sudden gust of wind got up. Two men had been delegated to hold closed the neck of the balloon so as to prevent the escape of gas, but as they wrestled with the wind-beaten silk pipe, it tore from bottom to top opening up one whole side of *Erin Go Bragh*. Word was then put about that Hampton was in the habit of puffing, and had only a limited number of successful flights to his credit, though he himself claimed a record in excess of eighty ascents. With matters taking this nasty turn, Hampton placed the whole affair in the hands of a committee, chaired by the Mayor, and gave an undertaking to 'give satisfaction' as soon as his balloon was repaired.[17]

Brown received a more sanguine report of events, with Hampton recording that the public bore the failures patiently. At the same time some of the tension under which he found himself was manifest in his snappy advice to Brown concerning the Montgolfier project. 'I would advise you to mind what you are doing respecting the fire-balloon'. And regarding the book on aerostation, 'don't be in a hurry to publish it yet'.[18] Brown took these rather negative responses well, and wrote to Hampton assuring him that he wasn't the only one who'd failed to get aloft to the disappointment of the public, and cheerfully, but tactlessly, gave a list of recent aeronautical failures, including a report of one fatality.[19]

Back in Cork Hampton had repaired his torn balloon and the committee set the date for the ascent as Wednesday 1 August. All proceeds from the event were to be handed over to the Cork Dispensary, to be used for the relief of the increasing number of famine victims.[20] This time there were no serious mishaps, but Mrs Hampton had to leave the car before there was enough buoyancy experienced to raise Hampton himself together with Mr Francis Jennings and Lieutenant Andrews of the 26th Cameronians. *The Cork Examiner* carried a rather sardonic account of events in its Friday edition, entitled, *How the Balloon did Ascend:*

> 'Yerra, didn't I tell ye it was all a humbug?' said a man near us, who had evidently made up his mind that aeronauts were the greatest public Diddlers unhanged – 'True to you, Mick,' replied his friend, between whom and himself there seemed to be a marvelous [sic] coincidence of opinion. 'He'll wait for the Queen,' said the third, in a tone of bitter mockery. 'He doesn't like to leave the Cork boys,' suggested a fourth. 'He didn't ate his supper,' suggested a fifth. In fact the balloon was fast losing its popularity, in consequence of its apparent disinclination to go ahead. The cause of

> the delay arose from too great a weight in the car, which had to be relieved of one of the sitters, Mrs Hampton ...
>
> We have now vindicated the reputation of Mr Hampton by duly recording how he got up; and we leave it to himself to tell how he got down.

And this he did in a report to the *Cork Examiner*, citing balloon capacity and dimensions, and also recording that the flight took fifty-five minutes and terminated with a descent at Cobh. Mrs Hampton, though displaced from her seat in the car, was not left without her moment of public notoriety. Queen Victoria and Prince Albert were due to visit Ireland at that time, and were scheduled to make landfall at Cobh, which they did at 10pm on the night of 2 August. The following day, as the Queen and her party stepped ashore at Columbine Quay, cannon were fired to signal her imminent arrival at Cork – within the day to be renamed Queenstown.[21] Hampton had manufactured three large unmanned gas balloons, each ornately decorated and inscribed in turn with the names, *Victoria*, *Albert*, and *Prince of Wales*. It was Mrs Hampton who now gave the signal for these balloons to be released, and they 'ascended steadily and gracefully amid the acclamation of several hundreds who witnessed the ascent. The feature was one of the prettiest in the day's proceedings'.[22] Hampton sent note of these latest successes to Brown.

News of Hampton's turn of good fortune seemed to have spurred Brown to attend once again to his balloon-building project, since he wrote back to Hampton, now in Limerick, announcing that he had managed to acquire a backer who would undertake to finance the construction of the balloon and engage Brown as the aeronaut.

What he did not reveal at the time was the identity of his backer: one Alfred Ormond, a would-be theatrical entrepreneur married to the actress Waverley Scott. Ormond had already tried to make his mark locally by staging a number of galas at the Botanical Gardens, Leeds, but had met with only limited success.

* * *

It is August 1849:

'You see, my dear, for all the guff he gave me about aerostation being a mucky business, Russum is still at it!' Charles turns from the window. It is a bright summer's day, and if one looks up towards Monkbridge Road, the clutter of the building development going on all around is not at all evident. Eliza is seated at the small sewing table, darning socks.

'I'm sorry, Charles. What did you say?' She pulls the sock from her hand and putting down the darning needle, sits composed to listen to her husband.

'Russum,' he repeats. 'He is flying still. Why, he even told me he will not let Green get the better of him. Indeed, Mister Hampton knows of Green, and has warned me he is like as not to steal any good idea he comes across.' Eliza cocks her head in puzzlement. 'But that is all bye the bye,' continues Charles. 'The question now is, how am I to proceed, so that I may have my own balloon.'

'But, Charles, had we not agreed ...' She hesitates. 'Agreed, that is, to leave matters lie till we are more settled here, and until Annie is a little older?'

Charles turns away from his wife and back to the window. For a moment or two he stares blankly at the net curtains. It is not clear if he is deep in thought or composing his temper. He turns and walks to the empty fireplace, drawing from his coat pocket a sheet of paper.

'Only a small amount is required,' Charles says, as much by way of a plea as a statement. He lays the paper on the sewing table. Eliza ignores it. She picks up the sock and slips it again on her hand. 'I am writing to Mister Ormond. He, I know, is looking for a venture that will give him a good financial return.' Charles takes up the paper and reads: '*I have made several improvements to the balloon ... and am anxious to meet a gentleman who will assist me in carrying out my design*', Charles is picking his way through the draft letter. '*A small sum only would be required from you – about £30 or £40 – and if the parties are honorable to each other, it could not fail to make a fortune for both.*'[23]

'And you really believe this Mister Ormond would be prepared to venture £30 on your say-so?' Eliza's tone is menacing.

'But I would provide my skills. I can build the balloon and repay his investment from the profits.' There is no mistaking Charles' tone now. He is pleading with his wife. She turns from her darning and looks at him from under lowered eyebrows. He turns quickly away, and blurts out: 'I have studied aerostation for many years. We cannot fail.'

'And apart from you time and energy in building the balloon, what money must *we* invest?' Challenges Eliza.

'I could venture £10,' suggests Charles.

'We cannot afford to dispose of £10 on nothing more than hot air,' snaps Eliza. She throws the sock into the sewing-box, and stands facing her husband.

'Eliza, my dear,' Charles takes his wife's hands in his. 'My love, what if – what if I were to meet Mister Ormond and merely discuss particulars with him?' Eliza pulls her hands free, but Charles grasps them again, and continues. 'And I shall make you the promise, the most solemn promise, that whatever the outcome of any arrangement ...' he shakes Eliza's hands as he emphasises each word, 'I shall not give up my situation.'

There is silence for a long moment. Eliza leaves her hands in Charles'.

'That is your solemn promise?'

'It is.' Charles' voice is firm.

'Very well then, I agree.'

* * *

Within a month the deal between Brown and Ormond had turned to dust. Ormond was unable to meet his debts. He had it given out that he had died of cholera, and then fled to Bavaria, where his 'widow' later joined him.[24] Brown was once again left with a partially constructed balloon and no prospects of being able himself to fund its completion. Hampton, now convinced that Brown was serious in his attempt to become an aviator, and wishing to minimise his impact as a competitor, Hampton took the opportunity to suggest that he would consider a partnership, but needed anyone joining him to invest £100. This was a sum totally beyond Brown's means, and he realised too that availability of half that amount would have allowed him to complete the construction of his own balloon.

Charles Henry Brown had many calls on his limited finances. He had been married for some time, though this was not something he had immediately intimated to his mentor, Hampton. It was only when his correspondence with Hampton was some four months old that Brown, for the first time, made mention of being married, though curiously, he did not reveal that he had a daughter, Annie.

On 5 September 1847 Brown had married Eliza Hartley at Saint Peter's Church, Leeds, and leaving his parents' home in Canning Street, had moved with his bride to a house in the expanding suburb of Meanwood. Meanwood Road was located near the old clay pits to the north of Leeds town centre. Leeds was undergoing a period of expansion. Many of the fields which had stood on either side of Meanwood Road had been sold off and turned into housing developments, which were crushed and crowded, being constricted by the former paddock's shape. This problem was augmented by the developers' desire to get the greatest value possible from their investment.[25]

In Ireland Hampton continued through the country with a degree of success. He had a flight from Limerick and a failure at Clonmel, and then moved on to Dublin. Along the way he gave the occasional lecture on aerostation, illustrated by the use of model balloons and miniature parachutes.[26] He continued to correspond with Brown, and solicitously inquired as to the manner in which he intended to construct his balloon, choosing to forget that he had already been told it would be a craft build on the Montgolfier principle. But Brown, willing to accommodate, and no doubt flattered by Hampton's interest, wrote back: 'As you seem desirous that I should give you some explanation of my plan for a balloon, I will at once proceed to do so', and he stressed, 'with respect to my being afraid of your taking advantage of any communication I might make that would be likely to be useful to you – on the contrary I should only be too happy if I could render you a service'. He then set out details of his intended modifications to the original Montgolfier design. His ideas included the fitting of a valve to the craft, a lid device on the brazier. He then turned around Hampton's earlier proposal and suggested that he might like to invest in the fire-balloon, the total cost of the machine calculated not to exceed £50.[27] He was beginning to see himself potentially as adept as his mentor. Hampton was quick to detect the tone of renewed enthusiasm in Brown's letter and was just as quick to caution. 'You say you have a wife,' he wrote, 'and that you hold a situation in the Post Office. Let me beg you to consider two things – the first – is your wife agreeable to your becoming an Aeronaut and to give up your situation for that of the precarious life of a public man?'[28]

Yes, Mrs Brown was, eventually, agreeable to her husband becoming an aeronaut, but, no, he would not give up his position. He would have to make do as best he could with that situation.

Meanwhile, back in Ireland, the Hamptons took lodgings at number 23 Jervis Street, Dublin. This handsome street was once part of the estate of Sir Humphrey Jervis, Sheriff and Lord Mayor. It was situated on the north side of the Liffey, in an area slowly becoming fashionable. There Hampton set about laying up the equipment for the winter, prior to setting sail for England where he and his wife planned to visit the Browns at Christmas time before returning to London. He hoped to be able to come to some more certain arrangement with Brown about his acting as agent, and since Brown's deal with Ormond had collapsed he now had less reason to worry about that young man becoming an independent competitor. But before he had finished in Dublin, a number of matters arose which both commanded his attention and required Brown's assistance.

Notes

1 Lionel Rolt, *The Aeronauts: A History of Ballooning 1783–1903* (London: Longman) 130.
2 J.E. Hodgson, *The History of Aerostation in Great Britain* (London: Oxford University Press, 1924) 234.
3 Brown to Hampton, 18 June 1849, *Correspondence*: 14–16.
4 Hampton to Brown, 11 July 1849, *Correspondence*: 18.
5 F.E. Dixon, 'Ballooning in Dublin', *Dublin Historical Record.Vol.XIV*. (Dublin: Old Dublin Society. 1955): 10.
6 *Evening Packet* 7 June 1849 & 7 November 1850.
7 Rolt 131.
8 J.P. Campbell, 'Two Memorial Dublin Houses', *Dublin Historical Record Vol. II*. (Dublin: Old Dublin Society, 1939): 141.

9 A. O'Donnell, *The Rathmines Township – A Chronology and Guide to Sources of Information* (Dublin: Dublin Public Libraries, 1998): 18–19.
10 C. Maxwell, *Dublin Under the Georges* (Dublin: Gill & Macmillan, 1979) 139.
11 *Saunder's News Letter & Daily Advertiser* 6 June 1848: 3.
12 'Balloon Ascent', *Saunder's News Letter & Daily Advertiser* 22 June 1848: 3.
13 Dixon 10.
14 Hampton to Brown, 11 July 1849, *Correspondence*: 18.
15 Brown to Hampton, 16 July 1849, *Correspondence*: 19.
16 'Grand Balloon Ascent from Cork Corn Exchange', *Cork Examiner* 13 July 1849, sec. Public Notices: 1.
17 *Cork Examiner* 18 July 1849: 3.
18 Hampton to Brown, 26 July 1849, *Correspondence*: 22.
19 Brown to Hampton, 5 August 1849, *Correspondence*: 23.
20 'Ascent of the Balloon', *Cork Examiner* 27 July 1849: 1.
21 C. Woodham-Smith, *Queen Victoria: Her Life and Times*, Vol. I. (London: Hamish Hamilton, 1972) 295.
22 *Cork Examiner* 4 August 1849: 4.
23 Brown to Ormond, 25 August 1849, *Correspondence*: 26.
24 Footnote, *Correspondence*: 28.
25 Asa Briggs, *Victorian Cities* (Harmondsworth: Penguin Books, 1980) 29.
26 Dixon 10.
27 Brown to Hampton, 29 September 1849, *Correspondence*: 29.
28 Hampton to Brown, 8 October 1849, *Correspondence*: 30.

CHAPTER 4

Trouble

The *Illustrated London News*, 13 October 1849, carried a front-page item under the heading, 'British Adventure.' It was fulsome in its praise of the spirit of the British as 'messengers of civilisation' and as a race which had done more than almost any other people towards the extension of geographical and scientific knowledge. The article, after more effusiveness, then got to the point:

> The long absence of Sir John Franklin ... has excited in the public mind an affectionate and deep interest, amounting at last to a painful solicitude for his fate and that of the brave men who share his perils and glory.

The Admiralty was urged to mount one more rescue expedition to search for those foolhardy explorers who had set off in search of the North-West Passage over four years earlier. Further, Lady Franklin was offering a reward for information leading to the rescue of her husband and his crew, or for any news relating to the missing vessels, HMS *Erebus* and *Terror*.

Still in Dublin, Hampton wrote excitedly to Brown. He had, for some reason, a falling-out with the English-based aeronaut, Coxwell, and he now warned Charles to be careful in any further dealings he may have with that party, whose conduct he classed as shameful and deceitful. Brown had been so focused on his dealings with Hampton that he had allowed his correspondence with Coxwell to lapse, and the way things were progressing at Dublin and Bradford, he saw no reason to renew contact with that gentleman. But there were matters of greater import to which Hampton now called him to turn his attention. Hampton had a plan for the rescue of Sir John Franklin. It was his belief that a hot-air balloon would provide the ideal platform from which to scout for the ships, *Erebus* and *Terror*. However, he needed Brown's help. He wrote:

> As I am somewhat deficient in my pen, I would feel much obliged if you would put the enclosed [proposal] in form, by writing it plain – and where you can add, do so, both in words and suggestions or improvements.[1]

The curious thing was another aeronaut, George Bercher Gale, had already laid a similar proposition before the Admiralty, and Hampton knew this to be so, but now, with Brown's help, he intended to present his plan directly to the public – in Ireland and England. He would send a copy of the edited proposal as a letter to the Dublin *Freeman's Journal*, and, he asked, would Brown find as many newspapers as he could that would publish the plan? Brown was prepared to do all of this for Hampton. The revised letter duly appeared in the *Freeman's Journal*, but it had been pipped there by a letter from Gale outlining his own plan for finding Sir John Franklin and his crew, a plan not dissimilar to Hampton's. Undaunted, Brown went to work again on Hampton's original missive and a newly edited version was published by the London *Sun*. It was carried also in the local newspapers at Cheltenham and Wolverhampton. *The Limerick Chronicle* printed an abridged account, which Hampton observed, was 'not worth notice', but for all of that his plan was still being paraded before the public.[2]

A potential problem lurked behind the cover of all this activity. George Bercher Gale and Brown were on friendly terms. Gale had made a number of ascents from Leeds, and had allowed Brown aloft for his first flight a few years before, although this had been nothing more than a moored test ascent.[3] In addition he had undertaken to show Brown's aeronautical manuscript to a publisher of his acquaintance. But Hampton did not know any of this and Brown did not see fit to inform him.

True to his word, Hampton and his wife arrived at Leeds to spend some time at Christmas with the Browns. Hampton took the opportunity to look through Brown's work on aerostation, making suggestions for some changes and he even suggested a publisher – William Shoberl in London. Encouraged by this support, Brown quickly edited the text, and entrusted the manuscript to Hampton to take to London. Within a few days of Hampton having delivered the book, Brown received a very polite rejection from Shoberl, who explained that the work would not be a profitable undertaking 'in these times which are so very inauspicious to literary speculations'.[4] Shorbel gave the opinion that the book could be got out privately for about £300, but he cautioned against risking such an amount. It was an unnecessary warning, since Brown would have been pressed to manage £30, let alone £300. In a curious coincidence, Shoberl was just then undertaking the publication of a work by John Shillinglaw entitled, *A Narrative of Arctic Discovery with details of the measures adopted by Her Majesty's Government for the relief of the expedition under Sir John Franklin.*

Over the short time he had spent with the Browns at Christmas, Hampton had become even more enthusiastic about the fire-balloon and its role in the proposed rescue of Franklin's party. The original idea of making public his proposal for a Franklin Rescue attempt had not proved fruitful. There had been no groundswell of popular support for Hampton's plan. By chance, Gale had met Hampton in London and had indicated that he intended visiting Leeds. Hampton now became very secretive about the whole undertaking. He was worried that Gale would get wind of the full particulars of his revised plans with regard to the Franklin project, and the great improvements he intended to effect on the rescue balloon, so he asked Brown to pretend that the two of them did not know each other. Brown was prepared to go along with this subterfuge, though for his part he did not reveal the extent of his acquaintance with Gale.[5] In truth, Gale posed no threat to Hampton, since he was not really in a position to mount a rescue. He was in severe financial difficulty and was obliged to sell off some of his equipment, finally borrowing money from Brown! Brown, all the time, remained tight-lipped about his friendship with Hampton and the particulars of that man's plans for a Franklin rescue.[6] In reality Brown considered the rescue mission to be more of his devising than of Hampton's.

The Admiralty had, before the arrival of the new year, bestirred itself in regard to the Franklin affair and organised a further rescue attempt. Hampton, for his part, had once again written to Lady Franklin concerning his plans for a rescue, and now the Admiralty called on him to present his drawings to them at Admiralty House on 9 January. In fact, two Arctic discovery ships were due to depart the very next morning. The *Enterprise*, under the command of Captain Collinson, together with Commander McClure's *Investigator*, set off for the Behring Sea, via Cape Horn. It was reported that:

> the day previous had been spent testing the properties of various small balloons, which their inventors wished to be taken with the expedition for the purpose of being committed to the winds of heaven, in the hope that some of them might be wafted to the spot where Sir John Franklin and his comrades are haply pent up.[7]

Hampton had not been invited to give a demonstration. It was true that the release of small 'message' balloons formed part of his overall rescue strategy, but his plan was centred on the use of a piloted hot-air balloon. His belief that the use of such a balloon was still under favourable consideration was reinforced by the seemingly receptive mood of the Navy. By the end of the month he was able to report to Brown, 'they received my plans favourably, and I should have gone out with this last expedition had there been time for me to have got ready'.[8]

The combination of Hampton's optimistic reports and his own failure to get into print contributed to Brown's increasing mood of dissatisfaction with his personal situation. To Eliza's dismay, Charles began to speak of giving up his job with the Post Office. With the New Year now a month behind him, he wrote to Hampton, by now back in Dublin making preparations for a new flying season and a possible call to head to the Arctic, and told him of his state of mind.[9] Hampton responded by writing to his 'dear friend', advising him against any rash action with regard to leaving his employment and suggesting that Brown 'keep it up till you can better yourself'.[10] This was easy enough advice to give. Hampton was at the time temporarily lodging in Cork Street, in Dublin's Liberties, a less salubrious area than the gracious surrounds of Jervis Street, but comfortable nonetheless.[11]

Gale, Hampton's possible competitor in the Franklin undertaking, had arrived in Leeds and was in dire financial straits. He found himself being pursued by creditors in that city, as well as others in Bradford, Wakefield and Hull. With a wife and nine children to support, he had little hope of meeting his obligations unless he could gain some engagements to make a balloon ascent, though it was dangerously out of season. He had some experience in other fields of entertainment, but he had no props or co-performers to enable him to mount a variety turn, and he had already set about selling his aeronautical equipment. In this way he was able to make some provision to feed and lodge his family. His prospects of making any money in the provinces were bleak and he needed to return to London.

* * *

Charles Brown emerges from the Post Office after his day's work. It is six o'clock in the evening and the gas lamps glimmering around the City Square scarcely counter the darkness. Charles stands on the steps and opens his umbrella, seeking some protection from the slanting gust of rain and sleet. He is surprised when his free arm is grasped. He turns to his left and is immediately face to face with George Bercher Gale.

Gale is attired, as is his wont, in his imagined approximation of nautical garb. Having spent some time in the Coast Blockade Service in the north of Ireland, he is loath to forego that connection, or surrender his assumed rank of lieutenant.

'My dear Brown, I have startled you.'

'Mister Gale,' says Brown, forgetting in his confusion, Gale's attachment to his title. 'Indeed you have startled me. What are you doing here?'

'I was in London, but there was little there to which I could turn my hand, so I came north. Leeds has always been kind to me.' Rain is dripping from the peak of his sailor's cap and running in rivulets down the front of his brass-buttoned greatcoat. Charles extends his umbrella to cover Gale's head.

'Come back into the vestibule,' he invites. 'It is freezing here.'

'And very wet, Brown,' adds Gale wryly and unnecessarily.

There is still half an hour of business to run before the Post Office closes for the day and the late working counter staff are busily tallying the contents of their cash drawers. A few people are about in the hallway. Gale drops a sopping duffle bag from his shoulder, and brushes off the excess moisture from his coat. He now seems uncomfortable, and rubs his hands together, as if to restore heat and circulation.

'And have you had any success?' enquires Charles.

'Success? How my dear Brown?' asks Gale, his tone defensive.

'In gaining some suitable engagement.'

'Oh, I see what you mean! No, my dear Brown, alas, no.'

It is now Charles' turn to feel awkward. 'I heard you had some little misfortune in London,' he ventures, by way of bridging the moment. He immediately realises he may have said too much.

'Indeed, Brown, and who gave you such information?' Gale has turned down his collar and laid his cap atop the duffle bag.

'I cannot immediately recall,' lies Charles.

But Gale has taken the comment as relating to other things. 'In truth, I have suffered a disappointment,' he admits. 'I had proposed a scheme to assist with the rescue of Sir John Franklin, and indeed I had even demonstrated one of my devices to the authorities some weeks ago at Woolwich Dock.' He looks away, distracted, and runs a hand through thinning locks. 'But, I regret, I had no success. There may be another expedition – mid-year, perhaps.'

'Is it because of this matter that you have sought me out?' asks Brown anxiously. It is now Gale's turn to be untruthful.

'Why, Brown, this has been mere chance. But a fortuitous happening.' He pauses, and this allows him sufficient time to reflect just how unconvincing he sounds. 'But, I would inevitably have sought you out,' he hurries on, 'had we not so chanced on one another.'

There is another silence between the men.

'Can I ...?'

'Would you ...?' They speak simultaneously

'I beg your pardon, Lieutenant,' says Charles, remembering the courtesy of the contrived title, and extending a hand by way of invitation for Gale to continue.

'You will recall,' says Gale, gently massaging the bristle on his chin, 'you will recall, I have in the past extended a favour to you, here and there?'

'How could I forget? It was you who provided me with the opportunity to take my first flight.'

'Yes,' smiles Gale. He nods to himself and becomes more animated. 'A jaunt in a balloon *captif*, I grant, but a flight for all of that.' Charles smiles back at the sodden sailor. 'And, in another matter I made a promise, as I recollect,' continues Gale. 'Yes, a promise to recommend your literary work to a publisher friend of mine.' He taps Charles on the shoulder with a finger. 'And immediately upon my return to London I shall take up that matter. As a most urgent priority.'

'There is really no need ...' demurs Charles, but is interrupted.

'Not at all, my dear Brown,' says Gale. 'When back in London the matter shall have my attention, *instanter*.' Gale is now genuinely pleased with himself. But only for a moment. 'Ah! Now there is the rub.' He takes to stroking his chin again. 'When I return to London.' He turns away from Charles and leans against a writing desk, drumming his fingers on it. There is a protracted silence between the two men. Charles realises he must take the initiative.

'Can I help?' he asks, resolved to render the assistance for which he knows he will be asked, even though it will involve a hardship for himself – and Eliza and Annie.

'I am presently bereft of all funds,' confesses Gale. 'I have no means whereby to provide for the transport of myself and family back to the Metropolis.' Gale is conscious of sounding pedantic, but he feels this infuses his statement with a degree of dignity, and he is now going to beg. 'May I impose on you, my dear Brown, for the loan of sufficient monies to enable me to return myself and my family to London?'

Charles fumbles through the inside pocket of his overcoat. 'Would £2 be sufficient?' he inquires; and dreads to think how Eliza will react when she discovers his monetary recklessness.

'Oh, adequate, adequate,' laughs Gale, unable to contain himself. 'There are excursion fares at present, and with the children half price £2 will be adequate.'

Three crumpled 10/- notes, a crown and two halfcrown coins are handed over. Gale pockets the money and fumbles in his bag. 'I have here a small token. A mark of my gratitude.' He pulls out a small packet wrapped in newspaper. Opening it out he reveals a half morocco bound copy of Mason's *Aeronautica*. It is in good condition, but would have fetched little or nothing from a pawnbroker. He lays it open on the writing desk and taking a steel ball-nib pen, he writes on the front endpaper: *Presented to Charles Henry Browne by George B. Gale, aeronaut, February 28th 1850.*[12] Looking over his shoulder, Charles makes no comment on the misspelling of his surname. Gale presses the book into his hand. 'A token, a mere token.

Please accept it. And, of course, you shall be repaid in full for your generous loan.' Gale hurriedly secures his duffle bag and heads towards the Post Office door. He pauses and waves back at Charles, who is still standing, mouth slightly ajar, and the book held limply in his hand. 'When I am in funds,' calls Gale cheerfully. 'When I am in funds, my dear Brown.' And he goes into the night, with his duffle bobbing merrily on his shoulder.[13]

* * *

For the foreseeable future, there would be no threat to Hampton's Arctic rescue plans from Gale. Brown could not resist telling Hampton of his *tête-à-tête* with Gale, which he now viewed as not only pathetic but slightly amusing. Hampton failed to see any humour in the occurrence, and reiterated his caution that nothing of the rescue plans be let slip to Gale, or anyone of his ilk.[14]

Some weeks later, Brown himself had cause to be no longer particularly amused with the Gale incident. True to his word Gale had spoken to his publishing contact, Mr Dipple, who had out of hand declined to deal with Brown's manuscript. Gale conveyed this bad news in two lines of a letter.[15] It was the last Brown heard from the bold lieutenant.

As time went by, however, it became clear to Hampton that his earlier optimistic hope of being called upon by the Admiralty to help rescue Sir John Franklin was not well founded. Lady Franklin was suspicious that despite the movement at Admiralty now to send ships to the Artic region, the question of rescue for the officers and crew of *Erebus* and *Terror* was becoming secondary to the push to discover a northwest passage. She wrote to Hampton to tell him of her fears.[16] But she had not abandoned all hope and at the same time she also sent a letter to Benjamin Disraeli asking if he could exert some influence since 'the Admiralty have not thought fit to take upon themselves the responsibility of renewing the search for the missing Arctic Expedition'.[17] This was not strictly true. A four-ship fleet under Captain Horatio Austin was scheduled to set sail in May. Hampton was aware that the feisty Lady Franklin had previously organised her own expeditions to search for her missing husband and his crew, so he wrote to her again, renewing his offer of service, should she decide to fund a rescue mission of her own.

During this time the work of developing the hot-air balloon still continued. Hampton was testing Brown's fireproofing compound and the two were exchanging technical information, with Brown making suggestions as to the placement of the furnace and its general management. Brown hoped to travel to Dublin to help with the project, and he was encouraged in this by Hampton. In the meantime he continued writing up papers for Hampton, the latest being about 'Hampton's New Aerostatic Surveying Fire Balloon.'

This craft was cut from the same pattern as the *Erin Go Bragh* balloon, and would serve as the rescue balloon. Unfortunately, the fireproofing which Brown had proposed for the calico envelope was proving unsatisfactory, and the question of the ideal fuel with which to 'rarefy the air' so as to give lift was also problematic. A mixture of phosphate of ammonia and ammonia was finally settled on as give adequate protection against fire to the envelope, and a concoction of linseed oil and spirit of wine would serve as fuel. This was a mixture which Brown himself had used to get unmanned exhibition balloons aloft. With these technical problems resolved, all seemed to be going on satisfactorily.

Notes

1 Hampton to Brown, undated 1849, *Correspondence*: 34.
2 Hampton to Brown, 3 December 1849, *Correspondence*: 40.
3 *Leeds Intelligencer* 20 May 1848: 5 and footnote, *Correspondence*: 187.
4 Shoberl to Brown, 26 December 1849, *Correspondence*: 44.
5 Hampton to Brown, 19 February 1850, *Correspondence*: 51.
6 Hampton to Brown, 21 April 1850, *Correspondence*: 61.
7 'Departure of Artic Discovery Ships', *Argus* 3 May 1850.
8 Hampton to Brown, 28 January 1850, *Correspondence*: 47.
9 D. Bennett, *Encyclopaedia of Dublin* (Dublin: Gill & Macmillan Ltd, 1991) 45.
10 Hampton to Brown, 13 February 1850, *Correspondence*: 50.
11 Bennett 109.

12 Dominic Winter Book Auctions, 'Aeronautica', 12 April 2001. <http://www.thesaurus.co.uk-winter/>.

13 Gale's little conceit concerning his naval commission is exposed in his entry in the DNB – ed. L. Stephen, *Dictionary of National Biography*, vol. XX. (London: Smith, Elder & Co., 1889) 373. Gale did serve with the Coast Guard Service but never held a commission.

14 Hampton to Brown, 5 March 1850, *Correspondence*: 53.

15 Gale to Brown, 20 March 1850, *Correspondence*: 55.

16 Hampton to Brown, 24 March 1850, *Correspondence*: 57.

17 Lady Jane Franklin to Benjamin Disraeli, 30 January 1850, 'Disraeli letters', 10 October 2001. <http://wwwbodley.ox.ac.uk/guides/modpol/resources/disraeli/htm>.

CHAPTER 5

Bad Day at Dublin

The weather in Leeds continued bitter during the first months of 1850. Brown was unable to obtain leave from work at the Post Office to travel to Dublin. This added to his growing feelings of frustration, as he was most anxious to be directly involved in the construction of Hampton's balloon. It was, he believed, a craft that would incorporate a great many of his novel ideas, and Hampton could only rightly claim to have introduced the use of wooden toggles instead of the usual knotting, as a means of securing the car to the envelope. Brown still had the small satisfaction of editing and publishing the material sent over from Dublin which was aimed at boosting public interest in the Franklin rescue bid.

But all was not going well with that project. Lady Franklin informed Hampton that, in her opinion, the most that could now be expected from the Lords Commissioners was the use of small message balloons with notices attached, a system which had already been trialled at Woolwich Docks, to the satisfaction of the navy. Now Lady Franklin opined, correctly, that the management of such small balloons would be left to the officers on board any vessel sailing to the Arctic region.[1] Hampton was not deterred by this information, and set about experimenting, sending up many small balloons with waterproofed messages attached, asking the finders to let him know where they had come to earth. This curious conduct was duly reported in the Dublin newspapers.[2]

While the effective fireproofing of the new balloon had been achieved, there still remained the problem of generating enough heat to inflate the craft. Spirit of wine was the fuel Hampton had in mind to use. Naptha when set alight had an overpoweringly disagreeable smell, and alternative fuels such as turpentine and rosin gave off a great deal of smoke. In all of this Brown noted to his distress that the craft Hampton was promoting in pamphlets and flyers (which he, Brown, was editing) was of monstrous dimensions. Even with the most apt fuel it would prove difficult to achieve a reasonable lifting power. Brown informed Hampton of his fears. Chastened, Hampton admitted that he had no intention of building such a gigantic balloon and had exaggerated the size of the vessel in order to excite public interest. Puffery of this sort had always irritated Brown, as it ran contrary to his view that aerostation should be treated as a serious scientific undertaking rather than a form of public entertainment, but in this instance he held his counsel and continued with his task of writing up promotion material. Hampton now came up with an alternative notion to help with the promotion of the balloon: 'I think you had better name the balloon as follows, "Hampton's New Aerostatic Surveying Fire Balloon".'[3] His idea was that the introduction of the aspect of fire would excite public interest in the craft. This promotional stunt Brown also let pass without comment.

By April Hampton had still not finished the manufacture of the new balloon. The proposed fuel, too, was proving unsatisfactory. Hampton was about to conduct experiments with charcoal and 'other things'. He asked Brown, 'Should you think of anything likely to answer, pray let me know'. Brown became more anxious than ever to get over to Dublin and render practical assistance in the building of the balloon. Hampton, picking up on this mood of discontent and in an attempt to counter it, encouragingly wrote to Brown: 'My balloon and car are at your service at any time and place for you to have an ascent at any time you can get leave'.[4] This did little to alleviate Brown's agitation – he had come more and more to think of the craft as *his* balloon.

The fraught situation in Dublin, augmented by the constraints of working at the Post Office, with its rigid hours and less than generous pay and conditions, was leaving Brown more and more distressed. Desperate not to lose his technical adviser and editor, Hampton, in an attempt to cheer-up the younger man, took up on a project that had long obsessed Brown – getting into print. Gale's sponsorship had proved a disappointment, but now Hampton, who for all his protestations to the contrary was no slouch with the pen, resuscitated the idea with a suggestion that they go into partnership for the publication of an aerostatic magazine. Already much impressed by Coxwell's little magazine from many years before,

Brown was enthused by the idea. It was agreed by the pair that when the new balloon had been trialled they would get together and make plans for the magazine's publication.[5]

Still the project in Dublin dragged on. Brown came up with a formula for the fuel: a combination of spirit of wine and linseed oil which he had used with great success in his unmanned balloons. But, having experimented with the new mix Hampton was not satisfied. While the fuel gave off sufficient heat, he claimed it was unstable, spat out boiling oil and would in time ruin the balloon.

The stress and nagging discontent finally became too much for Brown, and despite his promise to Eliza, at the end of April he handed in his notice to the Post Office. His hours of work had been long, leaving him with little time to attend to the Dublin business and his own balloon-making. And soon there would be an extra mouth to feed. Eliza was due to have another child before the end of the year, and the cost of living in Leeds was proving too much to be met out of his post office pay. The family would move to Bradford. Now buoyed by the prospect of getting across to Dublin, he was not slow in letting Hampton know of his changed circumstances, nor could he resist writing to his occasional correspondent, Doctor MacSweeny in Cork, and telling him of the news.[6] He confided that ill health had been a factor in prompting his decision. The doctor wrote back, expressing his sympathy: 'I was sorry to find you had been ill. You must take care of yourself. Aerostation cannot afford to lose you now ... You must live to see the triumph of aerostation, for triumph it assuredly will'.[7] And the good doctor felicitously advised regularity of living, rural walks and a large well-ventilated bedroom as a panacea for Brown's ills.

The move from Leeds to Bradford was not an entirely rash undertaking. Brown had secured for himself a clerical position with a firm of solicitors and was able to get reasonable lodgings in Harker's Houses, on the High Street in Bradford.

Even with the move and his changed employment, circumstances still did not afford Brown the opportunity of getting to Dublin. And work on the balloon was still agonisingly slow. By an accident in its manufacture – a number of mismeasurements – the false boast of the balloon's vastness was being rendered a truth. This meant more lifting power would be essential, and a suitable fuel had still not been devised. A desperate Hampton wrote, 'I shall not have the Balloon complete till the end of June, I fear. I am in great anxiety about it'.[8] He had every reason to be anxious, as events elsewhere threatened to thwart his plans.

In London a group of speculators had set about mounting their own Arctic rescue bid. An advertisement appeared in the London *Times*, headed 'Rescue of Sir John Franklin', calling on 'the rich and chivalrous' to contribute to the cost of constructing a flying machine capable of travelling at one hundred miles an hour, with which to affect the rescue. The estimated cost of construction was £8,000. The only other private rescue publically contemplated at that time was being mounted by Sir John Ross with funding from the Hudson Bay Company and public donations: no mention had been made of linking balloons or other aeronautical devices with either attempt.[9]

There was a renewed flurry of activity in Dublin. The balloon was finally nearing completion, and had again undergone a re-naming. Now it was being touted as 'Mr Hampton's *New Arctic Surveying Balloon*'. Brown suggested that a valve be added to the top of the craft, to facilitate a quick deflation. Hampton had been having second thoughts about actually taking to the air in a type of machine with which he was not entirely familiar. The site chosen for the ascent was the Rotunda Gardens, and while familiar with the venue, Hampton was all too conscious of the proximity of Dublin Bay should the wind be blowing east, and the Irish Sea, if the breezes were to veer north or south. As a precaution against landing in the water in what was essentially a paper craft, he opted to make a captive ascent, with the balloon firmly tethered to the ground by stout cord.[10]

Finally a date was set for the balloon's ascent. Handbills were distributed through the city centre and the *Freeman's Journal* on 6 July 1850 proclaimed on page one: 'Projected plan for the discovery of Sir John Franklin – The experimental trial of Mr Hampton's *New Arctic Surveying Balloon*.' The newspaper reported that the celebrated aeronaut was pleased to announce the completion of his new and improved fire balloon, and that the project had already been received with approbation by the Lords of the Admiralty and Lady Franklin. And as a gloss to the occasion, Mr Hampton further revealed:

> on the present occasion [he] will also send up several Messenger Balloons inflated with gas and rarefied air, showing the method of distributing dispatches, so long spoken of by the public press, this plan being a further improvement in this particular, also showing that the packets will be so protected as to remain quite uninjured should they even fall into the sea.

The experimental ascent and release of the Messenger Balloons was to take place on Thursday 11 July.

The weather then took a turn for the worse. Rain and high wind caused the postponement of the ascent till the following Thursday, 18 July. Hampton seemed blessed by the arrival in the city of Hamilton's touring panorama. This would stir up any lagging interest. Already noted in London, Bath and Exeter, the grand moving panorama of the Arctic region featured 'The Search for Sir John Franklin'. It was to show in the Concert Room of the Rotunda Gardens on Monday 22 July. The problem would be, should Hampton fail to get his balloon aloft this time on the due date, the panorama could well take his audience, and satisfy their curiosity over things Arctic in general, and Sir John Franklin in particular.

Back in Bradford, Brown had received the copy of the final promotional bill that Hampton wished him to edit. To his chagrin he saw that no mention had been made of his contribution to the whole project. All the plans were his own, except for the mode of inflation, and Hampton had kept him in the dark as to the fuel he intended to use. Brown quickly wrote off a sharp letter to Hampton: 'If the papers are published in the style they are written, the public will be informed that the whole has been designed by Mr Hampton, and not by myself ... I certainly have no objection to your making use of any plans of mine, if you will only give me credit for having projected them – this is but reasonable and fair'.[11]

Hampton, however, was far from being convinced of the reasonableness of Brown's claim, nor was he prepared to concede its fairness. 'When you take upon yourself to tell me you are the real projector, and that all the plans I have adopted are yours, I must tell you in return that you are not entitled to anything', Hampton snapped back. 'You, like many more, have read much about balloons, and you then flatter yourself you know more than the practical aeronaut ... but you yourself must be well aware you could never manage a balloon – even could you do so, you would be a better friend to your wife and family to have nothing to do with ballooning'. With these lines Hampton wrote *finis* to his previously enthusiastic relationship with Brown, and it would be months before the men established a new and cautious contact.[12]

Seemingly undeterred by Hampton's advice to leave off his pursuit of ballooning, Brown recommenced a vigorous correspondence with Dr MacSweeny in Cork and also wrote to John Wise, an American aeronaut who had just published a work on ballooning.[13] In his book, Wise had commented on Hampton's pioneering parachute descents. Brown thought it politic to inform him of his personal opinion concerning the man: 'He is very unfit for his profession – he has not the confidence in his balloon that you appear to have; and he says that the conduct of Mr Graham and Lieut. Gale, two of our professional aeronauts, is quite shameful, in allowing their wives to go up with balloons'.[14]

In Dublin Hampton continued apace with his preparations for the ascent. The weather remained inclement during the week leading up to the scheduled demonstration, so Hampton set about making the best of what held the prospect of being a bad day. He organised a number of bands to play during the preparation, inflation and ascent of the *Arctic Surveying Balloon*. Colonel Dennie of the 71st Highlanders and Colonel Furlong of the 43rd Regiment of Foot sent their bandsmen into the Gardens, and the boys from the Hibernian School came in from the Phoenix Park with their fifes, drums and brass instruments. Hampton, in a public relations gesture, invited the hundred or so pupils from the National Deaf and Dumb Institute to attend at the event free of charge.[15]

The day for the ascent, Thursday 18 July, arrived less wet but just as blustery as the previous week. In spite of this Hampton went through the various stages of preparing the balloon. With the balloon suspended, flaccid between two uprights, he set about applying a final coating of the fireproofing solution to the interior. The brazier was lit, using a combustible mixture, the formula for which he had kept to himself. The balloon was finally inflated and presented a gigantic spectacle, standing 'seventy hands high and forty-five feet in diameter'.[16] But that is as far as the event went. The wind was just too strong to risk allowing *The Arctic Surveying*

Balloon aloft. That was the last Dublin saw of the fire balloon. A week later Hamilton's panorama, 'The Search for Sir John Franklin', was drawing crowds to the Rotunda Concert Room.

Notes

1 Hampton to Brown, 24 March 1850, *Correspondence*: 56.
2 F.E. Dixon, 'Ballooning in Dublin', *Dublin Historical Record XI.* (Dublin: Old Dublin Society, 1958) 10. This stratagem had been used by Hampton in earlier years as a device to promote his ascents and to facilitate the recovery of any balloon downed at sea.
3 Hampton to Brown, 8 April 1850, *Correspondence*: 60.
4 Hampton to Brown, 21 April 1850, *Correspondence*: 61. The balloon here being offered is, of course, not the incomplete craft but the gas-powered *Erin go Bragh*.
5 Hampton to Brown, 5 May 1850, *Correspondence*: 67. For all his protestations to the contrary Hampton was able to pen a good lecture and about this time the publishers, Underwood, Eden Quay, Dublin, put out a little book, *Practical Hints on Pyrotechny and the Art of Making Balloons, for the use of Amateurs*. No author is credited, but there is a mention in the text of the manner of cutting a balloon pattern, which is 'the result of the experience of a friend'. Hampton is the most likely person to have provided this sort of guidance.
6 Brown to MacSweeny, 29 May 1850, *Correspondence*: 70.
7 MacSweeny to Brown, 31 May 1850, *Correspondence*: 72.
8 Hampton to Brown, 25 May 1850, *Correspondence*: 67.
9 This rescue attempt was lead by Ross, commanding the vessels, *Felix* and *Mary*.
10 Brown to MacSweeny, 6 June 1850, *Correspondence*: 74.
11 Brown to Hampton, 8 July 1850, *Correspondence*: 84.
12 Hampton to Brown, 13 July 1850, *Correspondence*: 85 & 86.
13 J. Wise, *A System of Aeronautics, Comprehending its Earliest Investigations, and Modern Practice and Art. Designed as a history for the common reader, and a guide to the student of the art* (Philadelphia: J.A. Seel, 1850).
14 Brown to Wise, 4 August 1850, *Correspondence*: 94.
15 *Freeman's Journal* 18 July 1850: 2.
16 *Freeman's* 2.

CHAPTER 6

Bradford Fiasco

Quietly pleased at Hampton's failure in Dublin, and ascribing it to a faulty fuel mixture and unwarranted alterations to *his* design for the balloon, Brown continued to complain of Hampton in his letters to MacSweeny and Wise, and no doubt to any locals who would listen. He told Wise of the failure of the *Arctic Surveying Balloon*, but avoided any mention of the extent to which the craft had been of his own design. The American, Wise, was an exciting correspondent. He called himself a Citizen of the World and proclaimed that 'the human family is entitled to the grand and glorious prerogative of Aerial locomotion – surely intelligent beings should be as much so as buzzards'. In turn Brown sent off his plans for a double opening valve as well as a drawing of his proposed parachute. Wise wrote back that he would thereafter use the valve and assured Brown that he would experiment with the parachute and should it prove to his satisfaction, 'I will notice it at the time as your invention'.[1] MacSweeny too was making his contribution to buoying Brown's spirits. He espoused a scheme which he termed 'Balloon Ways', that would see craft move from place to place along a network of guide ropes suspended on poles. He asked Brown to outline the proposal to Wise the next time he wrote, since, he believed, 'Balloon Ways, if once tried in the "go ahead" country, America, would soon spread'.[2] Excited by all of this, Brown continued with his collecting of aeronautical articles and cuttings and wrote letters to various papers on the subject of ballooning.[3] He was keen to re-establish contact with his correspondent of times past, Henry Coxwell, but that gentleman was in Europe, and having considerable success with his aerostatic undertakings, so Brown was forced to bide his time.

With the year coming to a close, Brown now turned his mind again to the pursuit of practical ballooning. He resolved to take to the air. The balloon which he had been constructing before he left Leeds was no more than half completed and had been left in storage. It was designed as a hot air balloon, and many of the ideas Brown hoped to incorporate in it had been passed on to Hampton to use in his *Arctic Surveying Balloon*. He had created a special fireplace and valve, and perfected to his satisfaction a fireproofing solution, the details of which he had also given to Hampton. Repeated attempts by Brown to find a financial backer for his balloon had failed.

By now Eliza had given birth to their first son, Frederick, and the subsequent call on the family's financial resources precluded Brown from making any outlay for materials to complete the building of his balloon. Just before Christmas, there arrived a letter from Hampton, by that time back in London. It was a strange document, part conciliatory, part pugnacious. Hampton invited Brown to visit him in London at any time through to next summer, and then by comparing correspondence Hampton believed it could be established who was in the wrong.[4] There was no immediate response forthcoming from Brown. He mulled over the matter as Christmas came and went. Then, well into January of the following year, he penned a letter to Hampton making no reference to his earlier offer of a meeting, but instead asking:

> If you are still in possession of the fire-balloon and wish to part with it, will you please inform me on what terms you will dispose of it – or if you are disposed to let it out for one experiment will you please state your terms.

He carefully enclosed stamps to cover the return postage. Hampton wrote by return, demanding to know if Brown thought he was in such penury as to be unable to afford postage on his mail. But he was not going to let an opportunity to turn a profit pass, insult or no insult. If Brown meant business, he had better then visit London and inspect the *Arctic Surveying Balloon* for himself, ascertain its condition, and then proceed with the deal, if he was prepared to bargain. Then, as a gratuitous aside, he commented:

> As for what I said to you in one of my letters to you respecting your incapacity to manage a balloon, [it] was only in friendship to you. Knowing that you are crippled

> in your arms; but if you think different it is not for me to say anything more about the subject.[5]

This curious reference to Brown being 'crippled in the arms' had indeed been made before by Hampton, but he was the only person who raised the subject. Russum made no mention of it in his attempt to dissuade Brown from a career in aerostation, made years before in Leeds. Nor did Brown respond to the comment when he wrote back and confirmed his interest in the balloon, requesting some indication as to the asking price. He expressed the view that the craft would be no use to Hampton as he was unlikely ever to get it off the ground, since he had been unable to concoct a combustible mixture powerful enough to impart a sufficient lift. Without further ado, Hampton announced that he would take £125 for the balloon. Brown decided it would be worth his travelling to London. He wrote to Wise telling him of his intention and hinted that some sort of reconciliation was in the offing. Wise quickly responded, offering his congratulations that the earlier dispute was to be resolved 'with coffee, without the pistols'.[6]

It was mid February before Brown went to London to view the balloon. Having seen it and noted its condition, he considered he would be better set to build his own vehicle. However, the meeting brought about an apparent reconciliation with Hampton, and Brown was content with the illusion. Hampton, anxious to reacquire the services of a competent editor, provided Brown with the name of a budding aeronaut who could possibly join in a partnership.[7] Brown returned to Bradford reinforced in his resolve to become an aeronaut. Again he wrote to Wise telling him of his plans and also revealing that, as far as he was concerned, the rift still continued between himself and Hampton. Wise was philosophical and perhaps by then a degree uninterested in the whole affair. 'You can get along without Hampton as well as he can get along without you', he opined, and he added by way of encouragement, 'I am really rejoiced to hear that you have determined to go into the business of Ballooning practically'.[8] Brown took one more shot at Hampton before letting that matter rest. 'I now see through his selfishness', he wrote to Wise. 'He is afraid of my coming into the field'.[9]

At the beginning of April 1851, Brown was still working feverishly to get his fire balloon ready, for he had committed himself to an Easter Monday deadline. He approached the Bradford Cricket Club and asked to use their grounds as the point from which to make his ascent. The cricket ground was located to the south of the city on Horton Road, right next to Little Horton Village and the Melbourne Alms House. It was an ideal site set on high ground and a flight of a mile in any direction, except to the northeast, would take a balloon clear of the city centre and into lightly populated areas. The ground was enclosed by a low stone wall and otherwise free of obstacles, though this advantage was off-set by the fact that there were no spectator stands. Still, Brown considered the location would suit his needs admirably. The club secretary, Mr Dawson, was keen on the project as he imagined there was a profit to be made from the gate takings. He summoned a club meeting for 7 April at the Rawson Arms, and 'earnestly requested' the attendance of the members.[10]

When the meeting was held, the members too were enthused with the idea of their grounds being the venue for the novel ascent of a hot-air balloon, and cheerfully resolved that Brown be given the go-ahead for his venture. It was known, too, that the Mayor, William Rand, was in favour of the project, figuring that it would be a feather in Bradford's cap to have a successful fire balloon ascent from the town.

On Thursday, 17 April the following advertisement appeared in the *Bradford Observer*:

> Bradford Cricket Ground – Aerostatic Novelty. C.H. Brown respectfully announces to the Inhabitants of Bradford and vicinity that he proposes to make a GRAND ASCENT with his BALLOON from the Cricket Ground, Great Horton Lane, on EASTER MONDAY, April 21, 1851 at four o'clock in the Afternoon.
>
> The Balloon is constructed on the original plan of Montgolfier, and will be inflated with Rarefied Air, and not with Gas. The Balloon has other peculiarities in addition to the curious method of inflation. It has no valve at the top, nor any network covering, as the Gas Balloons always have. Its capacity is about 25,000 cubic feet, and its ascending power is calculated at 600 pounds, exclusive of its own weight, which is near 300 pounds. Several pilot balloons will be launched during the afternoon, and at a Quarter before Four o'clock the Inflation of the Monster Balloon will be commenced. Persons desirous of witnessing this interesting process must be on the ground early, as the balloon will be filled in ten minutes.

> Doors open at Three o'clock – Admission 1s; children 6d. A Band of Music will attend during the afternoon. Schools admitted on Reduced Terms. Apply to Mr Jowett, Printer, Tyrrel Street.

Curiously, Brown was pitting his ascent against the attraction of Hegler's Circus Royal which was advertising an afternoon and evening performance on Easter Monday, with 'equestrian and gymnastic representations' to be held across town at the fairgrounds. Obviously he saw the novelty of his event as outweighing whatever attraction a circus might hold for the populace. It had been, after all, twenty-three years since Bradford had last witnessed a balloon ascent.[11]

There was still some work to be done on the balloon. Its size was increased by the insertion of extra gores, bringing its capacity up to 27,000 cubic feet. It was bereft of any ornamentation, Brown not having the time or funds which would allow him to paint the envelope in any way. The machine itself was constructed of yellow linen, lined with paper up to the crown, which was insulated with worsted stuff, a material which Brown believed to be flame-proof. And, despite his insistence that Hampton's *Arctic Surveying Balloon* be fitted with a large top-mounted valve, this craft was without any means of effecting a speedy deflation.

The weather over the Easter period had been uniformly good, and organisation at the cricket ground gave the appearance of being well under control, but on another front matters were beginning to take an unpleasant turn. Word had got about that Brown was the aeronaut Charles Brown from Sheffield. During the previous year, the Sheffield Charles Brown had had a number of disastrous attempts to get airborne at Halifax and Sheffield, and the rumour circulated that his impending attempt was a put-up job, calculated to bilk the spectators of their shillings.

Monday arrived, a splendid day, and Brown together with his wife was early in attendance at the cricket ground. By noon all had been prepared for the inflation. Brown took himself off to have some refreshment, and while at a local hostelry overheard a number of men discussing the wagers they had placed against the balloon getting up. Quickly returning to the balloon and finding Sergeant Carter of the local police and a number of constables in attendance, he asked for their assistance in keeping the crowd away from the balloon. Even though he had followed MacSweeny's advice to seat ladies in the front rows of spectators, in order that they may be a check 'on some men who may wish to be intrusive', many people were milling around the balloon.[12] To compound the problems Brown was now facing William Russum had arrived on the scene accompanied by his younger brother John. Brown had fancied a certain intimacy between himself and the older man, established in part out of their jousting about the merits and demerits of ballooning. Now he found himself confronted by a bellicose Russum who, supported by his brother, and whether from spite, over-excitement or over-indulgence, loudly voiced doubts to the Mayor and other dignitaries in attendance as to the prospects of the balloon actually getting aloft – with or without Brown on board. The crowd immediately took up on this observation and began to taunt Brown.

In the midst of all this confusion, the crown of the balloon was raised a few feet by means of cords running through pulleys from two tall posts erected at either side of the machine, and the process of inflation was begun. For whatever reason John Russum now invited himself on board the *Monster Balloon*, and stepped into the car with Brown. All attempts to eject him failed, and with the fire beginning to do its work, Brown turned his attention to making the necessary pre-flight adjustments. The machine was now inflating quite quickly and there was a need to slack off the external supporting cords. Brown, making the best of a bad job, instructed John Russum to attend to the fire, which he did with enthusiasm, and Brown made to leave the car and attend to the anchorage, when the balloon suddenly collapsed.

Trapped under the fallen envelope, Brown and Russum were faced with the danger of either being smothered or burnt. Mercifully, they were dragged from the flaming wreckage by friends who attempted to hustle Brown out of the grounds by a side gate, as the already restive crowd was now becoming uncontrollable. As they sought an escape across the stone wall from the developing disaster, the small group was intercepted by a man who threatened to put a bullet through Brown. Clearly he had bet on the balloon successfully ascending. Brown was once again rescued by friends and thrust into a cab which sped him and his wife home. Meantime a full-scale riot broke out at the cricket ground.

Safe at home, Brown had some time to reflect. He convinced himself that the collapse of the balloon was not as the result of poor construction or mismanagement of the inflation, but rather as the result of a plot hatched by a group of men had been laying bets on the failure of the ascent. Brown believed that the envelope was first cut with a knife, and then ripped by a youth commissioned for that work by the punters.[13]

Back at the Cricket Ground the crowds who had stood outside the grounds, hoping to view the ascent without the burden of paying the admission price, now invaded the area, climbing over the low surrounding wall. Some attempted to souvenir still burning pieces of the balloon, while at the same time many other spectators were endeavouring to get out of the ground through the narrow laneway leading out to Little Horton. In the ensuing mêlée rocks were thrown at the police, and a deal of damage done to surrounding houses.

Afterwards Brown was besieged in his residence by well-wishers and others. Some, having witnessed the conflagration and riot, were convinced he had been burnt to death or torn to pieces by the rioters. Through this excited crowd strode three men on a mission. The Chief Constable, attended by two detectives, arrived at 33 Harkers Houses, and demanded that Brown attend the Mayor at eleven o'clock the next morning in the Court House.[14]

The local newspaper was a little slower off the mark than the civic authorities, and it was not until Thursday of that week that it managed to get its version of the Easter Monday debacle into print:

> The Bradford Cricket Ground was the centre of extraordinary attraction on Monday last. In consequence of the day being a general holiday, and still further on account of the weather being remarkably fine, the crowds of sight-seers and lovers of the marvellous and perilous were swollen to an extent altogether unprecedented. By advertisement, and placard, and still more by oral communication, it had been made known to the entire population of the town and neighbourhood that Mr C.H. Brown, a young man from the establishment of Messers. Thornton, Firth, Ramsden and Co., intended to hazard the danger of the most fickle of all elements, and to travel whithersoever the winds might carry him in a Montgolfier balloon. It is scarcely necessary to state, except for the sake of the many boys and girls who flocked to the Cricket Ground on Monday, that a Montgolfier balloon requires no gas, and is open at the

bottom. It is, in fact, an enormous *fire* balloon, and rises in the air, because in consequence of the blazing naptha or turpentine in the fire place the atmosphere within the balloon is rarefied by heat until it is vastly lighter than that which is elsewhere diffused. The lightness of the air in the balloon is caused by the fire, and the balloon rises because of the lightness of the air inside it. It was announced that the ascent would take place at 4 o'clock; but long before that time anxious crowds of persons might be seen wending their way from every quarter towards the appointed place. We should suppose, on a rough estimate, that about 2,000 persons were admitted by payment into the Cricket Ground; but this number was but a very small proportion of the crowds outside and surrounding. Little Horton Lane was a complete fair, and at the end of all the streets; on the tops of every wall; in the branches of every tree; and standing wedged together in compact masses wherever the ground presented the slightest elevation, nothing was to be seen but countless human heads all stretched forth in eager expectancy. In Great Horton Road in the neighbourhood of the Cricket Ground a similar display of the 'million' was observable. Round about the Cricket Ground in the several fields adjacent were other multitudes, gazing and watching, whilst the neighbouring roofs and windows contributed their share of the great aggregate. It has been said, and on good authority, that 30,000 persons were assembled on the occasion; we know, however, how easy it is to over-estimate numbers, and prefer putting it down to 20,000, which all will acknowledge to be sufficiently moderate. As the time drew near for the grand result, increased interest was occasionally manifest, and the deep hurrahs of the distant masses broke with strange effect upon the ear. About this time we forced our way as near as possible to the balloon, which was composed of strong unbleached cloth or light canvas, lined through with a tough and light paper, intended to prevent the escape of rarefied air through the outer material. Two poles had been fixed in the earth at opposite points of a circle made of ropes extended on posts, and from the tops of the two poles a cord went across, from the centre of which the balloon was suspended. A band of music was on the ground, of which, in charity, the least the better said; or perhaps the musicians were unwilling to divert attention from the monster *rara avis* that lay sprawling on the grass, or flapping its heavy white wings in the wind. About 5 o'clock Mr Brown and another person or two, crept under the balloon, the side of which touched the ground, entirely secluding the car, the stove, and hiding from public view the entire operation going on inside. It was a long time before the machine began to swell, and meanwhile, disappointed hopes found plaintive utterance, mingled with boastful self-gratulation on the part of those who previously predicted a failure. We were exceedingly entertained with the vocabulary employed with reference to the affair. Walking round the ring we heard one say, as he significantly shook his head – 'It's a case!' Another said, 'It's a dodge!' A third denounced the whole proceedings as 'a sell!' A fourth, with a good old-fashioned indignation said it was 'a hoax!'

> A fifth, dissatisfied with all the expressions of sentiment ventured to hint that it was 'a fetch!' whilst a bluff old gentleman protested, with a very red nose, that it was 'all humbug and gammon.' Whilst, however, all these little pleasant indications of popular opinion were being uttered in stage whispers and in 'strictest confidence,' the work of inflation was steadily going on. 'A watched pot never boils,' and it would seem that a watched balloon never fills, for when we had satisfied ourselves as to the feelings of 'a liberal and discerning public,' we turned again to the balloon and found it two-thirds full of rarefication. The state of affairs now began to alter, and Mr Brown, who was decidedly at a discount when the balloon would not fill, began as rapidly to rise with his machine. Unfortunately, however, the hopes of the public were only excited in order to a more complete disappointment. In a few minutes, clouds of dense black smoke issued from the rolling monster. These cleared away and hearts again beat light. Alas, more smoke, thicker, blacker, longer continued; then a blaze; and instead of a fire balloon in the clouds, with one man in it, there was a balloon on fire, driven by the winds across the field, amidst peals of laughter, and some thousands of men after it![15]

On the Tuesday Brown duly attended at the Town Hall and witnessed the Mayor and a bench of three other magistrates, convict a total of ten accused on various counts of assault on the police and public disorder arising from the affray at the Cricket Ground. That duty being attended to, the Mayor then set to interrogate Brown. Mindful of his audience, and the complaints of those residing near to the Cricket Ground, he expressed the hope that Brown could convince the public of the practicality of his scheme on scientific grounds, adding that no man should take upon himself the responsibility of gathering together in excess of 20,000 persons – he himself having been amongst that number – unless he was confident as to the outcome of the event.

* * *

(*An elevated magisterial bench is upstage left, while a dock is placed mid-stage right. A small table is mid-stage centre, littered about with legal books and papers. A* CLERK *sits behind this table, while the bench is occupied by four magistrates:* THE MAYOR, MR WILLIAM RAND, MR WILLIAM CHEESBOROUGH, MR SAMUEL LAYCOCK *and* MR ALFRED HARRIS. *Downstage left half a dozen people are seated.* CHARLES HENRY BROWN *is one of them. There is a rather bedraggled young man in the dock* – PICKLES BOLTON)

MAYOR: Well, you've heard the evidence against you. P.C. Jennings has said on oath you assaulted him while he had in custody Michael Jordan, that man having been arrested for assaulting Sergeant Carter. You were trying to assist the man Jordan escape from police custody. What do you say to that?

PICKLES: It were all very confused, Your Honour. There were smoke and suchlike everywhere.

MAYOR: I am Your Worship, not Your Honour! The question is quite simple. Did you attempt to rescue the prisoner Jordan?

PICKLES: I bumped into Constable Jennings.

MAYOR: Knocking him to the ground?

PICKLES: Well, he fell, yeh.

MAYOR: (*to* CLERK *of Court*) I can take that as an admission of guilt?

CLERK: I would believe so. Yes, Your Worship.

MAYOR: (*to* PICKLES, *with a quick glance left and right to fellow magistrates*) The Bench finds you guilty. Fined one pound, with expenses of eight shillings, in default of which you will serve fourteen days imprisonment at Wakefield Gaol.

(PICKLES *looks about in confusion*)

MAYOR: (*continuing*) That will do. Stand down.

(PICKLES *is motioned out of the dock by the* CLERK *of Court. Hesitantly he crosses stage to* EXIT P)

MAYOR: (*continuing*) Have we further matters? I have a mind to call on Mister Brown for an explanation when court business has been attended to.

CLERK: Mister Hailstone would like to address the Bench concerning damage suffered with respect to his property, attendant upon the riot at the Cricket Ground.

MAYOR: (*again glancing at his fellow magistrates*) Since the matter would seem to be directly related to the criminal proceedings which have occupied us this morning, it would be apposite to hear what Mister Hailstone has to say.

(*The* CLERK *goes to mid-stage* OP *and leads on* MISTER HAILSTONE *from the wing. He is a man of agricultural appearance. He is helped into the dock and the* CLERK *takes the bible in hand and appears about to administer the oath*)

MAYOR: (*continuing*) I would imagine we could dispense with the oath in this matter, do you not think?

CLERK: As Your Worship pleases.

MAYOR: Now, Mister Hailstone, what have you to say to the Court about yesterday's occurrence at the Cricket Ground?

HAILSTONE: (*pompously*) I have much to complain about in reference to the exhibition of the previous day. I was put to considerable expense by the breaking down of fences and the trampling of my grass.

MAYOR: You live where, Mister Hailstone?

HAILSTONE: Up by Little Horton, immediate adjacent to the ground; toward Shear Bridge, Your Honour.

MAYOR: Your Worship!

HAILSTONE: Who?

MAYOR: Let it pass. Please carry on, Mister Hailstone.

HAILSTONE: So, Your Honour, I have a tenant in the lodge there, and the mob frightened her terribly. She was afraid the crowd would pull down her house, they were in such a rage.

MAYOR: How are you so quickly aware of the extent and cost of the damage done to your property?

HAILSTONE: I have a farmer-man in my service and he has estimated the damage to the fences and lodge alone to be upward of five pounds.

MAYOR: That much?

HAILSTONE: Not to mention the damage to the grass from the mob trampling on it.

MAYOR: Quite so. Thank you, Mister Hailstone.

HAILSTONE: These exhibitions are an intolerable nuisance.

MAYOR: (*irritated*) I am sure they are. Thank you, Mister Hailstone.

HAILSTONE: I hope Your Honours will make some remark upon them.

MAYOR: Yes, yes, Mister Hailstone. That will be attended to, straight away.

HAILSTONE: I am pleased to hear that, Your Honour.

(*With a deep reverential bow toward the Bench,* HAILSTONE *departs across stage to* EXIT P)

MAYOR: I can only express my deep concern that such conduct should have occurred in Bradford. It has been a perfectly disgraceful event. I hope that the penalties meted out today would prove a salutary warning to all those who would have a mind to create a disturbance at any future event of this kind.

CHEESBOROUGH: (*seated stage left on the Bench, bestirs himself*) The conduct of the police was admirable. They acted with perfect good temper and forbearance. (*He sits back, as if exhausted*)

MAYOR: (*somewhat taken aback by the sudden interjection*) Indeed. And I trust Mister Brown will be able to satisfy us as to the scientific validity of his experiment, and that it was not just a personal pandering to a futile, idle notion. It would be criminal if the undertaking were to prove merely a means of depriving the public of money, honestly earned.

CHEESBOROUGH: Has not Mister Brown failed in like endeavours before? Was there not a recent incident at Halifax where his craft failed to get aloft?

BROWN: (*jumping up*) That was not me, Your Worship. I am Charles Henry Brown. The other Brown is a Sheffield man.

CHEESBOROUGH: Are you not from Sheffield?

BROWN: No, I am not that Brown.

CLERK: Silence in court!

MAYOR: I believe that as an act of public duty, Mister Brown, you should establish that your undertaking was a practical, scientific … er … thing.

BROWN: Shall I speak from here?

MAYOR: No, take the stand. (*motions to the dock*)

(BROWN *crosses to the dock, bowing to the Bench when mid-stage centre. The* CLERK *of Court waves him on*)

MAYOR: (*continuing*) I must inquire of you, what was the nature of the experimentation you carried out, prior to this exhibition?

BROWN: (*pause*) None, Your Worship.

(*The whole Bench of four Magistrates becomes animated, as do the spectators*)

BROWN: (*continuing*) This type of balloon has been flown many times. The very first ascent was made in such a machine, in France.

CHEESBOROUGH: Are we to entrust ourselves to French contrivances?

BROWN: Mister William Russum, the Leeds aeronaut, of whom you will no doubt know, he has made nine ascents in a balloon of this type.

MAYOR: That being so, how do you now explain your failure?

BROWN: I had heard that many people were wagering the balloon would not get off the ground.

CHEESBOROUGH: That was a wager well won!

(*Laughter from the spectators*)

BROWN: (*defensively*) I believe that someone with an interest in the betting had a hole cut in my balloon and that was the main cause of the failure.

MAYOR: I was present, and it seemed to me that the balloon being consumed by flames had more to do with its failure than did any hole.

(*Again there is laughter*)

CLERK: Silence in court!

MAYOR: Could it be, Mister Brown, that you yourself had a wager on your own failure. As an insurance, perhaps?

BROWN: I must protest at such a suggestion. I have lost money on the venture. The balloon cost me twenty pounds, and the takings at the gate came to (*takes a piece of paper from his coat pocket and reads*) thirty-one pounds twelve shilling and seven pence. With the cost of fuel and the hire of the grounds I am four pounds out of pocket. And I have lost my balloon.

MAYOR: All very plausible, I am sure. But while I will not go so far as to declare this had been a decided hoax on your part, Mister Brown, there is to my way of looking at it, very little indication that there was a *bona fide* intention to take to the air. And I think, Sir, that you will find that to be the commonly prevailing opinion. You occasioned a gathering of a crowd, estimated at thirty thousand souls. You have a great responsibility in such an instance to ensure your conduct … your experiment … does not then furnish any excuse for disaffected persons to cause a disturbance. I will say no more.

(*There are murmurs of 'hear, hear' from the Bench and general movement from the spectators*)[16]

Notes

Despite Russum's attempt to dissuade Brown from involvement in aerostation and his belittling of Brown's attempt to go aloft at the Cricket Ground, he and Brown continued on relatively good terms and when from time to time they met they would amicably discuss ballooning.

Brown was not the only aeronaut having trouble with hot air balloons at that time. There is a newspaper cutting in the Baden Powell Collection at the Royal Aeronautical Collection, London; referenced merely as *Giant Montgolfier: Surrey Zoological Gardens. (1852)* It reads:

> The balloon did not, however, ascend on Thursday evening as promised, although it was then inflated almost to its utmost dimensions. It had no buoyant power, or at least not power sufficient to carry up the car, stove, combustibles and aeronaut. The cause of the failure was the constructor's miscalculation of the size of the furnace; a larger furnace was made, but was found to be incomplete, accordingly the first furnace was used, in doing which, from its small size, it became red hot, and the car of the balloon twice took fire. The aeronaut would even then ascend – the proprietor of the Gardens preferred disappointing the many thousand assembled. The vexation of the people became uncontrollable, and the balloon was torn into a hundred pieces.

Brown was not alone in his failure, nor was the London crowd's reaction different from that in Bradford.

1 Wise to Brown, 10 September 1850, *Correspondence*: 102.
2 MacSweeny to Brown, 8 October 1850, *Correspondence*: 106.
3 An example: His letter to the *Builder*, 23 November 1850, (Vol 8 ed.: 561.) disputing the merits of using a steam engine to provide propulsion for a balloon.
4 Hampton to Brown, 12 December 1850, *Correspondence*: 115.
5 Hampton to Brown, 22 January 1851, *Correspondence*: 130.
6 Wise to Brown, 29 January 1851, *Correspondence*: 133.
7 'Captain' W.H Adams. Nothing came of this, though Adams later went on to fly professionally at the Cremorne Gardens, London. (Hampton to Brown, 2 March 1851, *Correspondence*: 138).
8 Wise to Brown, 16 March 1851, *Correspondence*: 140.
9 Brown to Wise, 4 April 185, *Correspondence*: 145.
10 *Bradford Observer* 3 April 1851: 1.
11 Brown to Wise, 4 April 185, *Correspondence*: 149.
12 MacSweeny to Brown, 12 April 1851, *Correspondence*: 151.
13 Brown to MacSweeny, 22 April 1851, *Correspondence*: 158.
14 Brown to MacSweeny, 22 April 1851, *Correspondence*: 159.
15 'Announced Balloon Ascent: The Balloon Burnt. Disturbances and Arrests.' *Bradford Observer* 24 April 1851: 5.
16 *Bradford Observer* 5.

CHAPTER 7

A New Beginning

Confounded and not a little confused, Brown and his few supporters left the Town Hall. From early on in the project, Mayor Rand had encouraged its completion, explaining that it would, like as not, bring prestige to Bradford. Now, on the heels of the balloon's failure, he was offering no support whatever; he had to all intents and purposes turned on Brown and marked him as a pariah. Brown sought some way to justify himself. Friends undertook to arrange a further meeting with the Mayor, but Brown needed to do something a little more immediate. For all the caustic tone employed by the *Bradford Observer* in reporting the destruction of the balloon, the editor had a soft spot for Brown and, unlike the Mayor, was convinced of his *bona fides*. A sympathetic paragraph had been appended to the newspaper report of the magisterial sitting called to deal with the rioters. In it, an anonymous reporter asserted that Brown had been following the accepted practices employed for a Montgolfier ascent. Brown was afforded an opportunity to explain the lack of drag ropes and grappling irons. This he did, stating that they would be useless in his craft, which relied on heat as the agent for ascent and descent. He was, however, careful not to comment on the lack of a top valve, a device he had insisted Hampton incorporate in the Dublin balloon.

Encouraged by this turn of events Brown took time to compose himself properly and he set about preparing a defence a little more thoughtful and structured than the one he had stumbled through before the magistrates. He committed this revised defence to paper and the editor of the *Observer* was prepared to give it space. On 24 April 1851 the following appeared on the paper's letter page:

> Sir, – Will you favour me with space to offer a word of explanation in reference to the accident which took place on Monday in the destruction of my balloon, which

> I feel to be due from me to the parties who favoured me with their attendance. There are two or three causes from which it is [*sic*] resulted, the principal of which was that the fire grate was not constructed as ordered, so that the fire could not be extinguished at will. The next was from the strong wind prevailing, though this was not absolutely preventive, had other matters been right. Another was not having a sufficiently organised force in the ring, required by the violence of the wind, by which means many persons were about acting injudiciously, and whose presence was detrimental. To those who supposed it was a trick to cheat people of their money, I have only to say that the expenses I have been at are much greater than my receipts, not taking into account my own labour in making the balloon.
>
> I have only to add my great regret at its failure, and to assure the public, through yourself, that, at an early opportunity, as soon as I am able to get another balloon prepared, an ascent will be attempted to which persons will be admitted, by ticket, gratuitously, and every endeavour will be made on my part to give satisfaction.

Even before the disastrous events of Easter Monday Brown had decided to throw up his position as clerk with Thornton, Firth Ramsden & Co. His plan was to make a few ascents at Bradford, which he believed would gain him a sufficient following so as to enable him to raise some capital. He would then return to Leeds and commence ballooning from there. He had gone so far as to advise some of his correspondents to in future write to him care of the Post Office, Leeds.[1] But now the failure of his ascent and the destruction of his *Monster Balloon* caused Brown to modify his plans, and although he resigned his position he decided to remain in Bradford.

The rift with Hampton had by this time been, to an extent, healed. Brown wrote to Hampton telling him of his mishap, hoping for some small degree of sympathy and perhaps a little support. Instead he received a note from London in which Hampton told him he was displaying doubly bad judgment: firstly in giving up his position and secondly in having commenced as an aeronaut using a fire balloon. He cuttingly remarked, 'had you been more communicative, I should have tried to persuade you from spending your money and time on such a balloon'.[2] This was galling indeed, since Hampton had been the one who had settled on a fire-balloon with which to undertake the *Arctic Surveying Balloon* project. And he had conveniently forgotten how prepared he had been to take £125 from Brown for the sale of the very same craft. On the other hand, Brown was encouraging letters from Wise, MacSweeny, and M. Dupuis Delcourt in

Paris. All these men in turn passed what they considered helpful hints to the flattered Brown. Wise, on principle, was prepared to give encouragement to anyone who wished to fly. He sent a diagram of a varnishing machine he had constructed, which, he claimed, would make the doping of the balloon fabric an easy task. Brown was quick to see the advantages of the process and resolved to use this technique in the preparation of his new balloon. Delcourt had recently been added to Brown's list of correspondents. He had been involved with aerostation since the 1830s and was in the process of forming the first ever aviation society, *The Aeronautical and Meteorological Society of France*.[3] Brown was impressed by the man's aeronautical credentials.

MacSweeny had an academic interest in aerostation, and a particular bent towards hot air balloons. He held the firm belief that it was just a matter of time before dirigibility was achieved. As for Delcourt, the French had always espoused the use of hot air for the inflation of balloons, and although there was in France a legislative prohibition of going aloft carrying a fire, many successful ascents had been made in Montgolfiers, which had been inflated by large furnaces on the ground. But ever the practical man, Delcourt thought that a gas-powered craft presented the best prospect for an immediate success. In England carbonated hydrogen was readily available and cotton with which to build an envelope was cheap. He recommended Brown concentrate on the construction of a gas-powered balloon.[4]

Having mulled over the matter briefly, and in the light of his disastrous experience at Bradfford, Brown decided that Delcourt's point was a good one. Gas was relatively cheap, and the balloon envelope could also be constructed at a reasonable cost, using local materials. He placed an order for three hundred and seventy square yards of muslin, and started to look around for someone to join him in a partnership.

Gaining a partner proved to be a difficult task. This was not very surprising in the light of Brown's current notoriety, and initially no one could be found to provide the capital that would have made the financial burden of manufacturing the proposed balloon just that little easier. But by early May he was able to write to Delcourt and tell him of his good fortune in obtaining a backer – a friend, who would help defray the expenses of the undertaking. The actual identity of this benefactor was not revealed, and Brown refers to him (or unlikely, her) simply as 'a friend'.[5]

With the promise of money now behind him Brown was able to take delivery of the material he had ordered. He had chosen pink and blue muslin with which to construct the craft. He regretted not being able to decorate his new creation, but his own funds were severely limited, and his backer was unwilling to provide money for the expensive frivolity of painting the balloon.

The practice amongst aeronauts was to emblazon their craft elaborately with images or slogans, so in the event of not being able to get aloft they were at least in a position to provide the spectators with some sort of spectacle, even if it were nothing more than an ornate tethered balloon. The new balloon would have pink and blue gores, and that would be that. This time Brown opted to name his balloon *Celeste*. It was to be capable of containing 18,000 cubic feet of gas. He announced he would make his ascent from Wakefield. This location, he hoped, was just far enough removed from Bradford to provide a cushion against the poor publicity his failure with the Montgolfier had engendered.

The day set for the maiden flight of the *Celeste* was Monday, 14 July 1851. The weather was most unfavourable, with a strong wind blowing, but Brown was resolved to carry on with the flight. Mindful of the riotous conduct of the last group of spectators he had disappointed, Brown attempted to inflate the craft, but in the process of being filled with gas, the *Celeste* was tugged by a violent burst of wind and the envelope was ripped. The attempt had to be abandoned. Brown quickly rescheduled the ascent for the Thursday 17th July. This time the weather did not present an obstacle, but the repair of the damaged balloon had been hasty and shoddy. With the *Celeste* only half full, leaking gas and showing no indication that it had achieved any degree of buoyancy, the fuel company baulked at providing a further supply of gas. The spectators became agitated and, threatened scenes similar to those at the Bradford fiasco. As a distraction, Brown cut away the car, and released the unladen balloon in free flight. It vanished into the clouds and came to earth hour and a half later at Glass Horton, eleven miles away from Wakefield. Up to that point luck had allowed Brown to make the best of a bad job, but now, the balloon having fallen into a field of haymakers, they lustily set upon the strange apparition that had fallen from heaven and totally destroyed it.[6]

Brown's new backer had now lost their investment, and the enthusiasts who had initially supported him at Bradford, and who may have been reconsidering their positions vis-à-vis aeronautic adventuring, were now thoroughly dissuaded from associating themselves with his unfortunate endeavours. Mayor Rand, the equivocal supporter of Brown's earlier scheme, had been replaced in the mayoral office by the newly elected Samuel Smith. Mayor Smith was a gentleman more intent on seeing his private dream of a culturally enriched Bradford coming to fruition. The erection of 'a great building', which he had commenced as a private undertaking, was not seeing its concrete realisation during his tenure as mayor, with the laying of the foundation stone of Saint Georges Hall taking place during September.[7] He had neither the time nor inclination to associate himself with such ephemeral matters as aerial ascents.

By the start of October, the onset of winter brought the season for ballooning to an end. It was a time of year when those involved in the business of aerostation usually set about critically assessing their positions and planning for the flying season to follow in the next year, normally commencing about April or May. Brown had only a tattered balloon to show for his efforts up to that time, and there was no clear prospect of any opportunity to re-capitalise the venture. On an honest assessment Brown had no prospects for the following season.

It was then he received a letter from an aeronaut who had also undertaken an evaluation of their situation and was intent on making some adjustments. Hugh Bell was not only an aeronaut but an inventor. He had taken out a patent on a cigar-shaped balloon in 1848. Promoted as a dirigible, it had unfortunately proved not to be steerable. MacSweeny had noted the design when it first appeared in the *Patent Journal*, and commented to Brown about it: 'You will be surprised by all the matters he claims as being his invention'. The craft employed hand powered screws and a rudder. All these devices had been tried in the past, and had failed. But Macsweeny was prepared to give Bell credit for having shown a good example to other aeronauts for his professional approach to the science.[8] At the Vauxhall Gardens, from which he normally operated, Bell had had a number of disappointments, and on at least one occasion he blamed his failure to get the 'locomotive balloon' aloft on 'bad gas', a claim not calculated to enamour him to the proprietors of that establishment.[9] He had other balloons besides his patented one and he now wrote to Brown proposing the sale of one of these:

> The season for ballooning having ended, I should feel inclined to part with one of my balloons at a very moderate price. Should you still feel inclined to follow the science, and are in want of a balloon, you will find the present opportunity an excellent one for purchasing. Should the price be beyond your individual means, I should think that you will find no difficulty in getting a partner or two.[10]

Of course Brown's problem was just that. He was no longer able to find anyone to support his undertaking. In the hope of being able to persuade Bell to offer some arrangement which would not require too great a capital outlay, Brown confessed his financial predicament to him, but stressed he was committed to following ballooning and parachuting as a profession.

The idea that Brown would involve himself in practical parachuting was a novel one. It was true he had developed a design for a 'double parachute', with one canopy atop another. The thought was that the rush of air from the lower canopy to the upper would slow the rate of descent. The drop would be made with the parachutist ensconced in a basket rather than strapped in a harness. It was usual for the primitive parachutist to have a partner who would operate the release mechanism, freeing the chute from the balloon that had taken it aloft, and who would then be responsible for getting the balloon safely to ground.[11] Bell, sniffing the prospect of sale, encouraged Brown along this line of thought. He sang the praises of the balloon he was offering for sale, and declared it eminently suitable for use by a parachutist. And it just so happened he was in a position to offer Brown a parachute. It was not of Brown's double canopy design and Bell admitted he had not yet finished the manufacture of the device, but when completed it would be 20 feet in diameter and, claimed Bell, would be built on a 'much stronger and safer principle than any that has yet been constructed'.[12] All this was to Brown's liking save that when he got down to making a costing he discovered the deal to be too rich. The price of the package Bell was offering, balloon and parachute, was £175. In an attempt to clinch the sale Bell was prepared to offer terms. If the full purchase price was not immediately to hand he would consider an arrangement where Brown paid £125 down and three instalments of £25, each to be made at unspecified times during the following ballooning season. Either arrangement was totally beyond Brown's financial capability, but he looked for some way to prevent the opportunity slipping away. He countered with an offer entailing a protracted repayment, but Bell was as cash-strapped as Brown. Bell then floated the idea of Brown making a number of parachute descents, for which he would receive a fee of £5 a drop, and in this way he could accumulate some working capital. But when the time of year was reckoned, with the likelihood of the weather being poor, even Bell himself realised that the scheme was impracticable. Who would be prepared to sponsor such an event during winter?[13] Nobody could be found to fund the undertaking and the idea was abandoned.

The aeronautical circle being the rather tight-knit group it was – even though no great love was lost between its various members – Hampton soon heard of Brown's negotiations and sought to arrange a meeting with him, but was snubbed. His cautionary letter to Brown after the Bradford disaster still rankled, and the friendship had never been firmly re-established.[14] Ten months later, in August 1852, Bell wrote again to Brown. He was still attempting to make a sale of his equipment. The parachute had never been completed, so the balloon alone was being offered at £135. And as far as he was concerned, Charles Henry Brown had not abandoned his endeavour to become a professional aeronaut.

Notes

1 Brown to Wise, 4 April 1851, *Correspondence*: 149.
2 Hampton to Brown, 3 May 1851, *Correspondence*: 161.
3 Ben Mackworth-Praed, *Aviation: The Pioneer Years* (London: Studio Editions, 1990) 55.
4 Delcourt to Brown, 8 April 1851, *Correspondence*: 150.

5 Brown to Delcourt, 9 May 1851, *Correspondence*: 161.

6 Footnote, *Correspondence*, 172.

7 Asa Briggs, *Victorian Cities* (Harmondsworth: Penguin, 1980) 155.

8 MacSweeny to Brown, 23 July 1850, *Correspondence*: 89.

9 *Morning Chronicle*, 26 July 1850.

10 Bell to Brown, 20 October 1851, *Correspondence*: 165.

11 As early as 1845 Brown had boasted to Coxwell about his new design for a parachute, and four years later described the apparatus in detail in a letter to Hampton, himself a pioneer parachutist. In fact the design was not novel. Robertson the aeronaut had supervised the use of the device at Vienna in 1804, when a pupil, Michaud, made the actual descent. Marion Fulgence *Wonderful Balloon Ascents or The Conquest of the Air*. Chapter VII. 27 July 2001. <http://www.backpage.com/bppks/wombat/Pert17>.

12 Bell to Brown, 27 October 1851, *Correspondence*: 166.

13 Bell to Brown, 5 November 1851, *Correspondence*: 167.

14 Hampton to Brown, 22 October 1851, *Correspondence*: 165.

CHAPTER 8

Royal Gardens

It was 1852 and the Brown family was back in Leeds. Number 1 Park Row was situated just across from the Post Office on City Square. It was a bustling thoroughfare with none of the rural calm that had made Meanwood Road so pleasant. The removal from Bradford back to his hometown had been made without that element of triumph with which Brown had earlier imagined it would be imbued. And it was true that his earlier plans lay as wrecked as his last balloons. Eliza was as patient as was possible, but she had now been brought to the edge of quiet despair. There had been so many setbacks for her husband, and so much tribulation for the family. Their finances were fraught, but Charles saw the move back to Leeds as the start of a new phrase in his aeronautic adventure. He still insisted that there were prospects for the man who would persist in the field of aerostation, and if that meant he must contrive to remain by placing less emphasis on matters scientific and attend more to the titivation of the public, through the furnishing of aerial novelties, so be it – he had already compromised in this regard, with his ill-fated partnership with Alfred Ormond.

Brown did not write to any of his associates, even though he received anxious inquiries from MacSweeny in Cork, one of which inquired as to whether or not he had Brown's correct address! Hugh Bell, however, had obtained his new address. He was still keen to dispose of some of his equipment. Would Brown be interested in purchasing a silk balloon, with a capacity of 12,000 cubic feet, with a spare car thrown into the deal? All this for the bargain price of £135.[1] Brown was not interested, and he did not bother replying to Bell.

* * *

'Charles, will you not consider asking your father for some help in this matter?' Eliza is tending to young Frederick as he perches at the kitchen table paddling away at a bowl of porridge. He is a slight child, even for a three-year old, and is not always in the best of health.

'My father will not provide any assistance as long as he believes I am intent on keeping to aerostation as a means of livelihood.' Charles is at the head of the table. Behind him the kitchen window is open onto a tiny back yard. He is puffing at his pipe. It is a small luxury he allows himself.

'Then ask Edwin, he will surely help,' pleads Eliza. 'He does not disapprove of your plans.'

'Plans!' Brown puffs out a cloud of smoke with an audible 'phuhh'. 'What plans? Ideas? Yes! A desire? Yes! But what have I to lay before anyone – before the public – at the moment? I have nothing to propose as a specific and nothing to show except newspaper scraps of my failures.'

'Charles, Charles, you frighten me when you talk like that.' Eliza will not look at him. She busily attends to wiping Frederick's porridge splattered face.

'Any road, I imagine Edwin is as pressed for cash as we are.' Charles tamps his pipe with his thumb and allows himself a slight smile. 'Though I doubt if big brother would allow as much.'

'Why don't you ask at the Post Office for your old job back?' asks Eliza. Her husband is not a man of violent emotions, but she knows there are times when he can be pushed too far, and any hint that he may be in the wrong can trigger off a long bout of black silence. But this is not such an instance. Brown stands and slowly moves towards the kitchen door. He turns back for a moment and gently tousles Frederick's hair.

'I've to see a man this afternoon. He can get me some clerking.' He slips his pipe into his pocket. 'Don't worry, lass. I'm not totally daft.'[2]

'I know that, Charles,' says Eliza, looking at him. 'I know that.'

'Aye.' He lifts an envelope from the table and taps it against the back of his hand. 'And when I get back, I'll drop a line to the good Doctor MacSweeny, and let him know we're still in the land of the living.'[3]

* * *

When he finally re-established contact with MacSweeny on 9 August 1853, Brown was able to inform him that he had once again 'obtained a good position'. Always anxious to be supportive and to proffer advice, MacSweeny suggested that Brown should 'always hold in mind the Scotch maxim, that there is more tact required in keeping a good place than in getting it'. He warned Brown against 'wounding the prejudices' of his new employers by 'over zeal' on the subject of aerostation.[4]

Brown had obtained a position as a clerk with the solicitors, Barr and Nelson, and while his wage was adequate he still needed more funds if he were to pursue his objective of becoming a fully-fledged aeronaut. Not being able to rely on his father for any financial support, and loathe to approach his brothers, Brown had contacted Thomas Clapham, and had successfully convinced him of the advantage of using his clerical and administrative skills. Clapham was the proprietor of a pleasure garden.

The Leeds Zoological and Botanical Gardens were situated at Headingly some miles north-west of Leeds. The Gardens had been opened in July 1840 under the management of the Botanical Society, and admitted 2,000 on the first day. But, the initial excitement over, the Gardens began a down-hill slide and eventually, eight years later, passed into the hands of an energetic showman, Thomas 'Rufus' Clapham. He set about reviving their fortunes and introduced firework displays, gymnasts, gala-days and balloon ascents. He also arranged for George Hudson's Leeds Northern Railway to run excursion trains to the Gardens, stopping at the grandly named 'Botanical Gardens Gate Halt'.[5] The balloon ascents became a feature at the Gardens, by then known as the Royal Gardens. During a holiday period an ascent could draw a crowd of as much as 30,000.[6]

When Brown returned to Leeds he had quickly made himself known to Clapham and suggested that, since he had many contacts in the field of aerostation, and excellent organisational skills, he would be the ideal candidate to attend to the booking of balloon attractions. He also puffed his expertise as a maker of small pilot balloons and boasted that, if necessary, he could actually pilot any balloon Clapham might care to purchase. Clapham already had engaged the aeronaut 'Captain' Stephen Chambers on a casual basis, but he was very happy to have Brown look after the administration of future bookings, also on a part-time basis. Most of this work Brown could attend to of an evening and at weekends. With a secure clerical job and the casual work for Clapham, things were now beginning to look up for

Brown, and quite soon he came close to getting an opportunity to stand by his boast and prove his worth as an aeronaut.[7]

'Captain' Chambers had agreed to make twenty ascents during the 1853 Summer Season, and began by employing the balloon *Prince of Wales*, a craft he had brought up from the Cremorne Gardens in London. After the first few successful flights, Chambers' confidence seemed to have deserted him. He had a number of failures and in all only made ten flights.

A charitable gala day had been arranged for the Royal Gardens for Monday, 19 September 1853. The *Leeds Mercury* announced the event: 'A Grand Gala will be held in these princely gardens, for the benefit of the Leeds General Infirmary – Grand balloon ascent by Captain Chambers.'[8] Chambers was scheduled to go aloft that evening in his balloon, *Victory*. Then, as the time for him to prepare for the flight drew near, for some reason he decided he was not going to fly. Clapham was away in London on business, and Brown sent a desperate telegram to Great Northern Railways in the hope of contacting him. The ploy was successful, and Clapham promptly telegraphed back instructions that Brown should have the *Victory* filled and take it up himself. Armed with this authority and escorted by some policemen, an excited Brown went to the gardens. Seeing him arrive with his escort, and guessing his intent, Chambers panicked at the prospect of having his thunder stolen and had the balloon quickly filled with gas. By then a strong wind had sprung up making conditions dangerous for flying, and providing a genuine excuse to abort the ascent, but, perversely, Chambers and a companion proceeded to take off. He was not going to let Brown so easily usurp his position as *the* aeronaut. Chambers' first attempt to take the wind-buffeted *Victory* aloft resulted in a short hop and a crash landing. In an effort to achieve greater buoyancy the passenger was ejected from the car. This solved the problem of lift, and though the balloon shot up to a great height it stayed aloft for less than seven minutes. This was a paltry flight, but enough to satisfy the non-critical spectators. More likely than not Chambers had been imbibing at one of the many refreshment kiosks in the gardens before finally taking to the air.[9] But, once again, Brown had been foiled in his endeavour to get airborne.

For the moment, Clapham was prepared to persevere with the unreliable Chambers. With another Grand Gala scheduled for 1 October, the services of the band of the 28th Regiment already procured and public notices posted boasting, 'a grand balloon ascent by Captain Chambers R.N., the most courageous and venerable Aeronaut in the World', it was to Clapham's way of thinking better to persevere with the devil he knew, rather than risk the piloting to the willing but accident-prone Brown.[10] The year reached its conclusion without Brown being again presented with an opportunity to fly.

By now wary of ballooning as a paying proposition, particularly in the light of Chamber's erratic performance during the previous season, Clapham was slow to commit himself to engaging another aeronaut for 1854. But something in the way of a novelty could be attractive. A night-time balloon ascent, perhaps? He asked Brown to think about the matter and make a suggestion.[11] And no, he would not consider Brown for the job. Anxious not to lose his standing at the Royal Gardens, Brown believed Hampton was the only aeronaut likely to make himself available at short notice, but he had been moving further and further apart from his erstwhile friend. He even managed to contrive an insult out of a simple request from Hampton for the return of any of his old newspaper cuttings that Brown might still have in his possession. He charged Hampton with accusing him of stealing the scraps. 'When I wrote to you, asking if you had any of the newspaper cuttings of my ascents, I did not mean to say that you intentionally keep [*sic*] them back', wrote a startled Hampton in reply.[12]

Now necessity brought about by a belated attempt at reconciliation. Brown wrote to Hampton inquiring as to the terms on which he would make one or two ascents from the Royal Gardens for galas to be held during the Easter period.[13] By return Hampton plaintively informed him that his balloons, stored away during the off season, had been destroyed by fire, and with the insurance company being tardy in settling his claim, he would not be flying until the following season though he would probably travel to Dublin with his panorama.[14] That was the last direct contact they had with each other. Within the year Hampton had retired from active ballooning.[15]

Having had his first approach fall flat, Brown now needed to find someone who could quickly substitute for Hampton and make balloon ascents from Leeds. He then recalled that Henry Coxwell had recently returned from Germany where he had been demonstrating the use of balloons for military observation and signalling.

Coxwell had been one of Brown's earliest correspondents. Brown was still in his early twenties when he first wrote to Coxwell, who was at the time publishing a small magazine titled, *The Balloon or Aerostatic Magazine*. This was an eight-page publication that appeared only at irregular intervals, and Brown became an avid would-be contributor. He received, however, an occasional notice in the publication: 'Mr C.H. Brown is thanked for his valuable communication'.[16] And again: 'We tender our special thanks to Mr C.H. Brown for the information he has afforded us'.[17] One item he desperately wished to have published related to the manufacture and use of a Montgolfier balloon, advocating it as the safest type of craft, but that too failed to make it into the magazine.

It was now time, Brown decided, to renew his contact with Coxwell. Clapham was agreeable to this suggestion, but he wanted to ensure he was getting his money's worth. So he decided that while Coxwell would be offered some engagements, there would also be the added attraction of Richard Hayes, the acrobat, performing on a trapeze suspended beneath the balloon. This would ensure a good turnout.

Richard Hayes was one of a daredevil breed of trapeze artists who made a living – and many times fell to their deaths – performing beneath balloons. Some acted as their own balloonists. Victor Verdalle, a Frenchman, performed a routine where he ascended on his own balloon without a car and hung by his heels from a cord while smoking a cigar.[18] Another Frenchman, Charles Grellon, using a Montgolfier, hung from ropes performing acrobatic manoeuvres.[19]

The proposal that he ascend with a person hanging from beneath his vehicle was not one calculated to please Coxwell. He was a man conscious of the need to keep up appearances, and was very committed to advancing the scientific position of aerostation, as he understood it. In the early editions of his magazine he commented approvingly that, 'many scientific men were engaged in schemes for the furtherance of aerostation'.[20] And he counted himself amongst that group. In the past economic necessity had obliged him to grudgingly involve himself in such events though he still objected to that sort of foolery. He was prepared to concede, however, that, 'if well done, it proves an immense draw'.[21] So Coxwell accepted the offered engagement at the Royal Gardens. However, arranging a date that suited all parties brought about protracted negotiations. Coxwell was not idle, and over Easter and Whitsun travelled to Germany to make more ascents for the benefit of the military. It was not until August that Clapham was able to post notices in the local papers:

> Balloon ascent by night – A grand gala is to take place in the Leeds Royal Gardens on Monday next. The entertainment will conclude at night by Professor Henry Coxwell ascending in his continental war balloon, the 'Sylph', accompanied by one or two other gentlemen. We have been informed that the ascent will be of a very beautiful and novel character.[22]

By this time Hayes had found himself other sponsors for his talents, and was not available to display his prowess suspended beneath Coxwell's car. As an alternative it was agreed between Clapham and Coxwell (and this time the arrangement was quickly reached) that the *Sylph* would take a rack of fireworks aloft and discharge them at a safe height, a performance perhaps a little more genteel than having a trapeze artist swinging below, but one which was considerably more hazardous with fireworks being discharged proximate to volatile gas. Brown was delighted at being invited by Coxwell to make the ascent with him. The resulting adventure provided him with opportunity to prove himself a literary adept. He wrote enthusiastically to the local newspaper:

> To the Editor of the Leeds *Intelligencer*.
>
> Sir,
>
> Having accompanied Mr Henry Coxwell, the celebrated continental Aeronaut, in his moonlight ascent from Leeds Royal Gardens, on Monday last, permit me to lay before your readers a few particulars of our charming trip. Mr Coxwell's balloon is capable of containing 30,000 cubic feet of gas, and when fully inflated, carry four or five persons. Unfortunately, owing to some accident at the Gasworks, the balloon, at the time announced for the ascent, was very flaccid, and would have required several thousand feet of gas to have filled it completely. Mr Coxwell having kindly offered me a seat in his car I gladly accepted the offer, and about half past nine in the evening got into the car and prepared for the ascent. As the ascent was announced

to take place amongst a blaze of brilliant colored [*sic*] fires, these were attached to a light wooden frame, which, on leaving the earth, was suspended immediately under the car, Mr Coxwell giving me strict orders to lower the frame on our overtopping the lofty trees around us.[23]

After the balloon had been illuminated all around by means of colored fires placed at a few yards distance, the match was applied to the fireworks beneath the car, the balloon was liberated, and we ascended calmly into the air. I gave my hat a few turns, and then commenced lowering the fireworks which, to Mr Coxwell and myself, presented a beautiful appearance. We were loudly cheered on rising, and could hear the shouting for some time. As the moon shone brilliantly, the shadow of the balloon appeared upon the ground in advance of us, and, in crossing the fields, shewed us clearly our rate of travelling. In a few minutes the fireworks ceased, and shortly afterwards Mr Coxwell informed me that, in his opinion, we were about 4,000 feet high, and that we should endeavour to keep the balloon in a horizontal course at that height. I kept my eyes fixed on the earth, whose surface, immediately below and behind us, appeared prettily carpeted, whilst in advance of us, in the valleys, a thick fog appeared. We observed in the direction we were going a village brilliantly illuminated with gas (apparently) which we afterwards learnt was Horsforth. On leaving the gardens, we proceeded in an almost direct line towards this village, but we soon found the balloon to incline to the right, proceeding very slowly. We had for some time ceased to hear sounds from the earth, but Mr Coxwell having found we were descending, threw out a little ballast, which we had no sooner done, than we heard voices, apparently directly under us, saying distinctly, 'There's the balloon!' we continued to descend, going directly over what seemed to be a large farm. We heard the voices of two females, who said, 'Come down,' repeatedly, and also added, 'we will help you.' We thanked them, bid them 'good night', and throwing out a little ballast, went nearly horizontally for some distance. Again, all was still; not a sound could be heard from below. Presently we saw a railway train crawling along, which in a few minutes came almost directly under us: we heard the whistle very plain; then the train disappeared, having entered the Bramhope tunnel, over which we were perpendicularly (sic) at the time. We also saw this train emerge from the other end of the tunnel. We now came down within a few hundred feet of the ground; not a sound could be heard; and we proceeded onward at pretty nearly the same elevation for some distance, occasionally remaining perfectly stationary. We passed over two large mounds of something of a slate colour, with which we were puzzled a little, not knowing what they were. We saw these mounds on our return to the gardens, and they appeared to us to have been thrown up in constructing the Bramhope tunnel. Mr Coxwell thought of effecting the descent a little further on, and pointed out a field which he thought looked favorable, if we could only avoid what appeared to be a tall oak in the centre. We were almost stationary; and after looking around a little, Mr Coxwell remarked that the tree was movable, and inquired if I recognised it. I looked and found it was the balloon shadow, which by this time had removed to the next field. We were much amused at this little deception. We reascended a little, as we could hear no sounds of any kind. Then, after proceeding a little further, Mr Coxwell called out, 'Where are we?' which was echoed very loudly. We heard a voice in reply, but could not distinguish what was said. I called out loudly, 'Where are we?' which was also followed by an echo and the words, 'Up in a balloon'. At this we laughed heartily. I then shouted out, 'What place is this?' and received the pleasant reply, 'Would you like to know?' Mr Coxwell called out, 'Shall we come down?' and received the answer, 'Yes, if you like.' Mr Coxwell then inquired, 'Is that a grass field' which was replied to by, 'No, it is a corn-field.' We had not yet caught a glimpse of our nether and humorous friend. We conversed with him for a short time, and presently saw and heard him walking through a corn field. We came down very low, so that the firework frame, which was ten to twelve yards below the car, trailed along the field. Our friend directed us to a suitable spot on which to alight, and we reascended a little. We disturbed a covey of partridges, and also some cattle, which ran about furiously. We came down in sight of a mansion, which I thought was Cookridge Hall, but our friend below, who still kept up with us, said we were far beyond Cookridge. We came down in a second field in front of the house and slid along the surface, frightening some cows, one of which looked fiercely at us, and did not seem inclined to move off till the balloon came almost close upon it, when it raised its tail and flew away rapidly. I threw out a little ballast, that we might bound over the hedge into the next field. We cleared the hedge: Mr Coxwell pulled the valve string smartly and after bumping slightly twice, we came down gently, the firework frame having caught in the hedge and anchored us. Two men came up, one of them being the person with whom we had conversed as we sailed through the air, who stated that he had followed us across the fields a mile and a half; the other was Mr Joseph England, gamekeeper to John Bastow Esq., of Greengates, Carlton, two miles beyond Horsforth. As we were within a few yards of the residence of the latter gentleman we requested Mr England to ask permission to empty the balloon, but finding the family had retired for the night and not wishing to disturb them, we accepted the offer of Mr England, who, with the other person, agreed to tow the balloon over the fields and fences to Mr England's farm, where we emptied the balloon with the assistance of six other persons, who came up half dressed, having turned out to see what they called the 'girt big thing.'

Our delightful and interesting excursion occupied an hour and a quarter, and we had travelled about five miles and a half. It was about quarter past eleven when we landed. By twelve o'clock the balloon was packed up and in the car and removed to Mr England's barn. Mr England treated us very hospitably, and for his and Mrs England's kindness we are very thankful. After refreshing ourselves, and resting a couple of hours, not being able to procure a conveyance, we resolved to return on foot to the Gardens, and arrived there about half past three in the morning. This

> terminated our nocturnal trip, the first that has ever been made in a balloon from Leeds. It was the 237th voyage of Mr Coxwell.
>
> Yours obediently Charles Henry Brown Leeds, Aug 9th. 1854.

Well satisfied with his literary effort, Brown dispatched his report to the newspaper, and was gratified to see it appear in full. He also broadcast the news of his flight to MacSweeny in Cork. The doctor, ever gracious, extended congratulations on his successful trip with Coxwell, and added the wry aside; 'it is a long time since I heard any other news from you … I knew not but that you had left Leeds'.[24] Brown admitted to MacSweeny that the flight had not affected him quite as he expected. But since many people had by now formed the opinion, consequent on his repeated failed ascents, that he was actually afraid to take to the air, he had purposely made something of a display of himself as the balloon went aloft.[25]

Brown had rescued his reputation as an aviator.

Notes

1 Bell to Brown, 3 August 1852, *Correspondence*: 168.

2 *Leeds Intelligencer*, 8 August 1854, reports Brown, 'Mr Nelson's clerk' as a passenger with Coxwell, in the *Sylph*. Barr and Nelson was a firm of solicitors located on Park Row. Wm. White, *White's 1853 Leeds & the Clothing Districts of Yorkshire* (1853; New York: A.M Kelly, 1969) n.p.

3 'The long time since I received a letter from you, made me suppose that you had left England', wrote MacSweeny. This was in consequence of previous letters of his having gone unanswered. 5 August 1853, *Correspondence*: 168.

4 MacSweeny to Brown, 12 August 1853, *Correspondence*: 170.

5 'The "Leisure Centre" with a bear pit', *Weekend Post* 28 June 1980: 14.

6 Alfred Mattison, *Chronicles of Leeds*, Vol I. (Leeds Public Library, unpublished MSS) Mf.942.75m435: 30.

7 Brown to Coxwell, 28 June 1854, *Correspondence*: 180.

8 *Leeds Mercury* 17 September 1853: 1.

9 Footnote, *Correspondence*: 172.

10 *Leeds Intelligencer* 24 September 1853: 1.
11 Clapham to Brown, 17 February 1854, *Correspondence*: 174.
12 Hampton to Brown, 12 December 1853, *Correspondence*: 174.
13 The letter was formal in tone, commencing, 'Sir, I am instructed ...' but it was the first contact the parties had had in three months. Brown to Hampton, 1 March 1854, *Correspondence*: 175.
14 Coxwell to Brown, 24 April 1854, *Correspondence*: 178.
15 J.E. Hodgson, *The History of Aeronautics in Great Britain* (London: Oxford University Press., 1924) 235.
16 H. Coxwell, 'Notice to correspondence', *The Balloon or Aerostatic Magazine* No.2 October 1845: 4.
17 H. Coxwell, 'Notice to correspondents', *The Balloon or Aerostatic Magazine* No. 3, November 1845: 4.
18 Brown to MacSweeny, 21 December 1849, *Correspondence*: 43.
19 Hampton to Brown, 24 March 1850, *Correspondence*: 57.
20 H. Coxwell, 'Editorial', *The Balloon or Aerostatic Magazine* No. 3, November 1845: 3.
21 Coxwell to Brown, 26 March 1854, *Correspondence*: 176. Richard Hayes had flown with Coxwell before and performed his trapeze routine suspended beneath his balloon at the Woolwich Gardens the previous summer. Coxwell did not care to be reminded of this lapse in decorum.
22 'Local and other news – Balloon ascent by night,' *Leeds Intelligencer* 5 August 1854: 5.
23 The framework, on which the fireworks were arranged, was only lowered from under the basket when the balloon had gained sufficient altitude.
24 MacSweeny to Brown, 19 August 1854, *Correspondence*: 186.
25 Brown to MacSweeny, 24 August 1854, *Correspondence*: 188. Brown chose to ignore his ascent at Headingly with Gale in May of 1848 in any reckoning he made of his flights, perhaps because it was a tethered ascent.

CHAPTER 9

Moving On

Another cause for Brown's enthusiasm at this time was the fact that, after the successful ascent at the Royal Gardens, Coxwell had asked him to manage the recovery, packing and dispatch to London of the *Sylph*. Brown was flattered by the aeronaut's trust.

Coxwell had devised a plan to interest the British military authorities in the use of balloons for reconnaissance and signalling. To Brown's way of thinking this was another opportunity to involve himself in scientific ballooning, and one which on the face of it offered better prospects of success than had Hampton's ill-fated Franklin venture. But as yet Coxwell had not in any way suggested that he would need Brown's assistance in forwarding this project.

Though he had managed to interest the German authorities directly, Coxwell for some reason felt obliged to take a rather more oblique approach with the British military and first engage the interest of the general public, very much in the same manner as had Hampton. He organised an ascent from the Surrey Zoological Gardens. According to the announcement he placed in the newspapers, spectators were to imagine his balloon as floating at a certain altitude over a beleaguered city or fortress. Communication with the ground was by way of signal flags, the defenders being the only ones with the key to the signal code, could be forewarned of the approach of the enemy and informed of their numbers and manner of deployment. Though cumbersome and far from efficient, the process involved such a theatrical display that it caught the imagination of the public and excited favourable comment in the press.[1]

Between trips to continental Europe, Coxwell was notionally engaged as the resident balloonist at the Royal Pavilion Gardens, Woolwich, but this was not a situation that provided sufficient remuneration to satisfy him. He had received what he termed 'cheap overtures' from several garden proprietors, but had turned these down. He asked Brown to make local inquiries. 'I have half a mind to try the Cloth Hall', he wrote. 'I wish you would enquire, specially the gas and all particulars'.[2] Brown did as requested. It was still early days in September, and the weather was holding – almost an Indian Summer.

Brown's activities as 'agent' on behalf of Coxwell did not go unnoticed by Clapham. He came to hear of attempts by Brown to procure engagements for Coxwell at Halifax and Pontefract. Clapham was angered by Brown's entrepreneurial initiative, but he held his peace. He had his own way of dealing with this subversive competition. Ignoring Coxwell he once again engaged the services of the erratic 'Captain' Chambers to provide ballooning entertainment at the Royal Gardens.[3] During late September, Chambers could manage no more than two indifferent ascents from the White Cloth Hall in Leeds. Disappointed, Clapham then reverted to his home venue the Royal Gardens and, in an attempt to spice up the program, went back to the idea of having a trapeze act associated with a balloon ascent. He engaged the Italian Brothers, aerialists who performed suspended below a balloon, and had William 'Bill' Green bring up the balloon *Victoria* for use in the act. This craft had previously flown successfully from the White Cloth Hall, but that had been back in 1849. The event was set for Monday 2 October 1854. William Green was called Bill to distinguish him from his father, a brother of the veteran aeronaut Charles Green. Charles Green had a certain standing in the public eye, having pioneered the use of coal gas for balloon inflation and flown his balloon at the coronation of George IV.[4] John Hampton had a more jaundiced opinion of Green and his brothers and had early cautioned Brown against any dealings with them. However, Hampton was out of favour with Brown, who was keen to meet Bill Green, a man not much older than himself, and who was, most important of all, a practical aeronaut.[5] It was true that William, Henry and George Green had been prepared to use their brother's reputation to advance their own standing as aeronauts and in turn their children were quite happy to adopt the same stratagem.

And so it was that Bill Green, with more gumption than skill, made his ascent with the *Victoria*. He carried aloft only one of the Italian Brothers, the balloon not having sufficient lifting power to allow the second acrobat to be taken aboard. Edouard Bonthellier was the brother who made the ascent. He immediately clambered under the car and commenced his performance. Even with the reduced load, the *Victoria* barely cleared the trees around the gardens. Desperately Green tossed overboard all the ballast, grappling ropes and Bonthellier's street clothes. The balloon came to ground at Bramham Park, fifteen miles from Headingly, and the unfortunate Bonthellier had to make his way back clad only in his trapeze tights.[6] Coxwell would not have approved. Nor did Rufus Clapham, who in his frustration turned on Brown and challenged him over his private dealing with Coxwell. The two had a first rate quarrel.[7]

Matters went from bad to worse for Brown. Clapham refused to give any engagements to Coxwell, and Brown was unable to generate any business. To confound matters further Mrs Coxwell fell ill, and Coxwell decided to make an early end to the year's flying and attend to his dentistry. This left Brown in a bind, but he busied himself in writing papers for Coxwell, who was still promoting the idea of using a balloon for military observation and signalling. Coxwell proclaimed this to be his contribution to the rescue of ballooning from the hands of showmen and mountebanks. It was his way of demonstrating that ballooning could have a military and scientific purpose.[8] There were, however, many who considered the whole undertaking 'humbug'.[9] During the brouhaha of the preceding months Eliza had quietly gone about the business of producing the Browns' third child, a rather sickly infant, Eugene.[10] Brown now attended to his family and his clerical position and hoped the rift between himself and Clapham would soon end, though he did nothing in particular to bring about a reconciliation.

By the beginning of the new year, 1855, the break between Brown and Clapham had to some extent healed itself, though each remained wary of the other. Coxwell, down in London, was busily working at his profession of dentist and maker of dentures, but he had not given up on aerostation and was, he admitted to Brown, only waiting for the beginning of the good weather and an improvement in his wife's health before he took to the air again. Nor had he given up on his obsession with ballooning as a military tactic. He asked Brown if he would make time to write to the newspapers, *The Times* or *The Mercury* being suggested, and enumerate and extol the benefits of 'Balloon Signals for the purposes of warfare' or alternatively give an 'impartial' critique on the application of balloons for observation. Brown was most intent on having his friend return to the air in Leeds, but he was certain Clapham would require something more spectacular of his aeronaut than the vigorous waving of flags by the balloon's pilot. Brown suggested to Coxwell that a balloon race might be well received at Leeds, but Coxwell ignored the proposal. And Clapham had not entirely forgiven Brown's carry-on with Coxwell the previous year, and at the Royal Gardens Chambers still continued in favour as resident aeronaut.

So the situation in Leeds seemed not to offer any great prospect for Coxwell. But Brown, loyal to his friend, continued to press Clapham to re-engage him. By February with no prospect of an engagement in sight Coxwell was becoming anxious. He still believed that an ascent featuring 'Balloon Signals for the purposes of warfare' would bring in the crowds.[11] It was then that a degree of lukewarm wheeling and dealing commenced between Clapham, Brown and Coxwell. Clapham considered Brown an excellent worker and did not wish to lose his services, but he was also a potential competitor. By having Brown continue his duties at the Gardens, Clapham was able to keep an eye on his activities. The relationship between the two men continued to be tense. In May Coxwell took the initiative and offered to make two ascents for £30. The erratic 'Captain' Chambers merrily undercut that offer. Clapham for his part then suggested an engagement at £12 for a night ascent. Coxwell in turn countered with a proposal for a day ascent at £20. This was declined. Coxwell, perhaps by now strapped for cash, offered one of his balloons for sale to the Royal Gardens, assuming no doubt that Clapham would use Brown to take it aloft. His asking price was £120. That was much too much for Clapham. By now it was mid-year, and Coxwell had not yet taken up a balloon, and offers were slow in coming from other centres. He was forced to admit to Brown, 'it does not appear that we are destined through Mr Clapham's propositions to come together … and you thus see another proof of his excessive selfishness'.[12] Brown, however, had not given up his quest to have Coxwell back at Leeds. Disappointed with his lack of success with the Royal Gardens Coxwell still managed, through his own efforts, to finally obtain a number of well-paid ascents from various London venues.

With his financial situation now stabilised, Coxwell had some time to look about, checking on events within the aeronautic circle. He constantly had his eye open for a bargain. Word got to him that a number of balloons had been seized under writ from an aeronaut in Manchester and were about to be sold off to defray that man's debts. He asked Brown to take an expenses-paid trip to the city and check out the state of the vehicles. Coxwell believed the asking price was in the region of £60, and he was prepared to pay that, should the condition of the crafts be satisfactory. Even closer to home an old balloon, *The Rainbow*, previously owned by the deceased Bristol aeronaut, William Wadman, had come on the market. Coxwell enquired; did Brown have particulars of that balloon amongst his voluminous files? Indeed he did. It was a jinxed craft, a very porous balloon, of 20,000 cubic feet capacity, and Wadman had been killed flying the machine.[13] Later Richard Green had used the balloon, only to have it come down over the Bristol Channel, and jumping from what he believed to be a sinking car, he drowned. Now the ill-fated vessel was on the market.[14] Following Brown's revelation concerning the history of the machine, Coxwell was, understandably, not interested in the craft, but he still wanted Brown to assess the Manchester balloons.

All this activity was revitalising to Brown. He set about making small balloons and other novelties that could be sold off to the various gardens he now intended to visit. Bellevue and Pomona were possible sites from which Coxwell could ascend. But ever practical, while touting for Coxwell, he also set about getting more engagements for Stephen Chambers, whom he now unblushingly referred to as 'Captain' Chambers, and whom he was promoting at the asking price of £25 per day ascent, a fee on par with that sought for Coxwell. But the undertaking was far from altruistic, since he was careful to extract a ten percent commission from the 'Captain'. There was some embarrassment when Brown discovered that at least one of the bookings he hoped to secure for his new client had brought him perilously close to running his man up against Coxwell. Robert Bowser of the Hull Zoological Gardens declined an offered engagement from Brown on behalf of Chambers, explaining that the gas supply at the venue was insufficient to inflate a balloon, and adding that he had also turned down an offer from Mr Coxwell for the very same reason.

It was mid-August when Brown took the trip to Manchester to check on the balloons being offered for sale. Coxwell had written a few letters urging him to make the trip and reminding him that Manchester's two pleasure gardens, Bellevue and Pomona, had in the past presented balloon ascents and may be interested in the prospect of another flight.

There was action on other fronts. Dr MacSweeny had written from Cork. He had been contacted by Hampton and the pair had turned to discussing things aeronautical. Although Hampton was by then primarily concerned with promoting his diorama of the Crimea War, he did remark on a book he intended to publish: an autobiographical piece about his life and times in aerostation. He had shown MacSweeny his notes and some of the prints he intended to use, but confessed to being at a loss to find someone who had the expertise to arrange for the publication of the document. If this was a hint to MacSweeny, he chose to ignore it and in turn suggested that Brown would be the man for such a job, provided some proper scale of remuneration could be arranged. Hampton asked MacSweeny to make the proposal on his behalf. This the doctor did, though he cautioned Brown that the task would be tiresome. 'You will have trouble with his notes ... to make out his meaning, as he writes a very illegible hand', he warned.[15] Brown, having dealt with the publisher Shoberl himself, had a good idea what was required in terms of layout and typeface. He accepted the commission. Mending the rift between himself and Hampton was not his primary goal, however, and he demanded that a proper arrangement for remuneration be made. For his part Hampton began to equivocate, realising the amount of work he would have to undertake to get the manuscript ready for publication, and then the cost involved in having Brown set up the book. Brown waited for his erstwhile friend to make up his mind.

Coxwell sent news from London that he was engaged in experimenting with a new invention and building a 'new machine'. He cautioned Brown to secrecy – an easy enough undertaking, since Coxwell remained totally silent as to the specifics of the device, though he boasted it would 'put all balloons in the background'.[16] Chambers, also down in London, was making but a slow recovery from injuries he had received when his balloon had made a very heavy landing. In his cavalier style, he had taken his balloon aloft without a car. Because it was insufficiently inflated he had perched on the hoop. When the poorly controlled balloon came to earth, Chambers had been thrown across the hoop and seriously injured his back.[17] Now, in the light of his slow recovery, he had decided to put his balloon, *Victory*, on the market. He asked Brown to find a buyer for it, naming £60 as the lowest price he would take, and at the same time offering Brown a commission of ten percent, should he manage to dispose of the craft. Brown was not successful. 'I am sorry to hear you had such a bad offer made', wrote Chambers after one inquiry, 'but whoever it was, they could not have been any judge at all'.[18] It was then that Brown suddenly realised this was an opportunity that would allow him to become involved in practical aerostation again. Chambers was obviously in no condition to continue with his aeronautical undertakings; leastways for the foreseeable future. Brown asked if he would perhaps be prepared to loan out his balloon. He was promptly rebuffed by the injured Chambers, but was invited to put in a bid for the purchase of *Victory*. In turn he declined, but countered with an offer to hire out the vessel. This proposition too was spurned by Chambers. Brown realised that he was only going to acquire the balloon through outright purchase. He made an offer of £50, with payment to be made on terms. Chambers was prepared to take £30 cash and the balance in three months' time, provided Clapham would act as surety for the outstanding amount. There was no way that Clapham would go along with such an arrangement and the deal foundered. So, for all the activity on his part, there was little Brown had achieved, and it now appeared that Hampton had become disenchanted with the idea of publishing his book. Overall it was not a good end to the year.

Notes

1 John MacKenzie Bacon, *Dominion of the Air* (London: J. Long, 1911) 103.

2 Coxwell to Brown, 15 September 1854, *Correspondence*: 190.

3 Chambers had no right to claim either the title Lieutenant or Captain, never having been in the armed services, but he adopted one or other of the ranks as the fancy took him.
4 Bacon 47.
5 Hampton to Brown, 14 August 1849, *Correspondence*: 25.
6 Footnote, *Correspondence*: 191.
7 Brown to Coxwell, 4 October 1854, *Correspondence*: 191.
8 J.E. Hodgson, *The History of Aerostatics in Great Britain* (London: Oxford University Press, 1924) 263.
9 Brothers to Brown, 22 June 1856, *Correspondence*: 232.
10 Eugene died at Melbourne 2 April 1875, the cause of death being recorded as 'heart disease and grand dropsy.' *Registry of Births, Deaths and Marriages*. Cert 6306. 1875.
11 Coxwell to Brown, 5 February 1855, *Correspondence*: 196.
12 Coxwell to Brown, 5 August 1855, *Correspondence*: 201.
13 Brown to Wise, 4 August 1850, *Correspondence*: 98.
14 Brown to Wise, 4 August 1850, *Correspondence*: 99.
15 MacSweeny to Brown, 3 January 1856, *Correspondence*: 214.
16 Coxwell to Brown, 19 October 1855, *Correspondence*: 207.
17 Coxwell to Brown, 19 October 1855, *Correspondence*: 206.
18 Chambers to Brown, 22 November 1855, *Correspondence*: 208.

CHAPTER 10

Something in the Air

The 1856 flying season had begun early for the Leeds Royal Gardens. Chambers was still incapacitated and trying to sell the *Victory*. Since the Gardens had no balloon Brown could not offer his own services as pilot. There was, however, some good news for Brown. Clapham announced he had decided to engage once again the redoubtable Coxwell. Brown was elated, as Coxwell had repeatedly offered to have him come aloft as his co-pilot.

Coxwell was booked to ascend on Easter Monday and Tuesday for a fee of £25. Well aware of the trouble Chambers and Green had experienced at the gardens, and additionally briefed by Brown on problems with inflation at the venue, Coxwell asked Brown to ensure there would be no trouble with the supply of gas. Delighted to have his friend return to Leeds, Brown quickly complied with the request. But the whole arrangement came close to falling through when Clapham, whom Coxwell regarded as a sharp dealer always on the look-out for number-one, attempted to have a Wednesday ascent tagged to the deal. It was not the prospect of an additional ascent that disturbed Coxwell, but he was angered at the suggestion that he undertake the ascent for a paltry £5 fee. 'The idea of such a thing!' exploded Coxwell. 'He must think me a fool.'[1] Brown quickly moved to play peacemaker and counselled Coxwell to let the matter rest.

The Easter gala was planned to be a splendid event. The balloon ascents notwithstanding, the main attraction was to be the presence of Lord Cardigan's charger, Ronald, which he had ridden at the Battle of Balaclava. Clapham had obtained the beast for exhibition and the services of its handler, Corporal Rigby, who undertook to ride him around the Gardens during the day. He had acquired this top attraction in return for a £50 donation to the Leeds Infirmary.[2] As his part in the entertainment

Coxwell was to make two ascents in the *Sylph*. Brown had done his job well and checked the local gas company. He was assured it could supply the 30,000 cubic feet needed to fill the balloon. As a precaution he hired two tailors to be in attendance and quickly deal with any holes or rents which might develop in the balloon's envelope, or worse, might be made on purpose by any malcontents or punters who had wagered against the *Sylph* taking to the air.[3] To top off this excitement all was now set for Brown to remind Coxwell of his promise of a place in the balloon.[4]

The two Easter ascents were a success. Every time it took to the air the *Sylph* stayed up for over half an hour, though the undertaking was not entirely without its misadventure. On the Tuesday Coxwell and Brown took a paying passenger on board; Jonas Proctor, a Pudsey clothier. The flight took them beyond Bradford at an altitude of 6,000 feet, though mainly through clouds. On landing the *Sylph*'s car received a severe shaking and Brown was struck on the head with the balloon's hoop, which left him senseless for a time. For all of his shaking-up, Brown finished the day a happy man.[5]

A peace treaty ending the Crimea War was signed in Paris on 30 March 1856, and Coxwell excitedly told Brown of his belief that, 'as peace is proclaimed, balloon ascents are likely to be more frequent'.[6] On this count he was not entirely correct. Balloon ascents were indeed seen by many organisations as being the ideal way to celebrate the victory over the Russians. But the intent was not often given realisation. Many of the proposed venues were unsuitable, there being no adequate gas supply to hand, and the expense was a deterrent to many. In the hopeful anticipation of a good season, Coxwell had started the manufacture of a new balloon. The as yet unnamed blue and yellow craft was to hold 32,000 cubic feet of gas. A little bigger than the *Sylph* it was not a large balloon when measured against some others being flown at the time. But Coxwell prided himself that its construction was such that it would have an excellent capacity to retain gas, and this would give him an advantage over his competitors. He invited Brown to be his companion on the craft's maiden flight. The undertaking seemed to have distracted him entirely from his earlier notion of building a craft of his own invention that would render balloons obsolete.

The increase in engagements the pair had anticipated as part of the celebrations to mark the peace, did not materialise. Brown and Coxwell agreed that the Whitsun holidays would be the ideal time for the new balloon's first flight. With this new goal in view, for all his effort and enthusiasm, Brown, however, did not manage to secure a firm booking for Coxwell. To his disappointment two gala committees at Bradford declined the offer of an ascent. He spread his inquiries as far afield as Derby and Nottingham, but without success. Finally, Brown was obliged to acknowledge that the season was slow. This view was confirmed by Coxwell who reported that even Simpson, the proprietor of the very popular Cremorne Gardens, had complained that attendances were falling off for balloon ascents.

Whitsun came and went and it seemed as if the ballooning season would be well into summer before Coxwell's new craft took to the air. It was then that a letter arrived from Mr Brothers of the Nottingham Arboretum Committee.

The Arboretum, designed by Sir Joseph Paxton, had opened in 1852.[7] Their Anniversary day had been set for Monday 16 June, and Brothers inquired of Brown if the aeronaut would be 'disengaged on that day, should we be inclined to place the engagement in your hands for a balloon ascent on that occasion'.[8] Coxwell, who had begun to wonder if he had misplaced his energy in the construction of a new balloon that seemed destined not to fly, was thrilled to hear of this likely engagement. 'One thing is certain,' he told Brown, 'you have established a character for perseverance.'[9]

Quickly Brown communicated the availability of his aeronaut to ascend on the Anniversary Day, and proffered his services. Brothers accepted, and asked also that some small balloons be supplied. These would keep the spectators amused while the large craft was being inflated. Brown was adept at making these balloons and he had also perfected the manufacture of what he was pleased to call his 'Colossal Figure'. This was a paper manikin some fourteen feet tall which when filled with gas or hot air would rise, waving its arms and nodding its head.[10] Terms were agreed and a contract signed:

> Agreement. I agree with the Chairman and Committee of the Arboretum to make an ascent in my 'Sylph' Balloon from the grounds of the Arboretum on Monday 16th

> June 1856 for the sum of twenty five pounds, at my own cost and risk, the Committee finding a sufficiency of gas and the requisite assistance during the inflation:
>
> Henry Coxwell
>
> I undertake to carry out the above agreement.
>
> Chas. Hy. Brown. May 16. 1856.

The use of the *Sylph* was a disappointment for Brown, as it forestalled his trip in the new balloon. The committee also wanted Coxwell to give a display of his war signals – at no additional cost. He was only too happy to oblige, as it gave a scientific colour to the ascent.

For his trouble in securing the ascent for Coxwell, Brown received a commission of five percent. By mid-year a steady flow of offers was coming in. The Fair Committee at Bradford, who earlier in the year turned down an opportunity to have an ascent, now wanted one from Peel Park on the 16 June. This clashed with Coxwell's undertaking to fly at Nottingham but Brown was able to convince the Bradford Committee that two ascents on the 17 and 18 June would be a better proposition. What was unusual about the arrangement for the Bradford Fair was that the ascents were to be made *captif*. It was understood by the organisers that Coxwell's new balloon would be used, and they wanted to hold a lottery for places in the car. Brown accepted the arrangement, though he knew this would deprive him of a chance to take a flight.

The situation had moved from famine to feast, with Clapham seeking Coxwell's services for two flights towards the end of June to be made at Halifax. The haggling over price commenced. Then a singularly unexpected element was thrown into the mix. Coxwell informed Brown that, if pressed to meet his commitments for the 16, 17 and 18 June, he would have another aeronaut substitute for him at Bradford, since he must personally attend at Nottingham to demonstrate his war signals. This alternative pilot was William Stewart, a London man who had been around the fringes of aerostation for some time having met up with Coxwell as early as 1849. As to his competency as a flyer, however, Coxwell felt obliged to admit, '[he] can hardly throw in life enough to satisfy employers and establish confidence'.[11] The arrangement was that Stewart would prepare the Bradford balloon and Coxwell would endeavour to get to the venue in time for the ascent, but should he fail, then Stewart would pilot the craft. The hurtful thing to Brown in all of this was the way in which he saw himself removed from the equation. Coxwell was well aware how hard Brown would take his exclusion from the car at Bradford, and the minimisation of his role in the balloon's preparation, but Stewart was an old acquaintance with more practical experience than Brown, who had not yet flown solo. In an effort to alleviate hurt, he increased Brown's commission to ten per cent.[12] Brown made no comment. At least he had arranged the start of Coxwell's flying season.

The ascent of the *Sylph* from the Arboretum at Nottingham was a success, though the war signals were considerably less impressive than the organisers had anticipated, being merely a display of various coloured flags. In the early development of the war balloon concept Coxwell had experimented with messenger pigeons, but they had not performed as efficiently as had been hoped. When trialled in London 'one knowing bird flew direct to the Stock Exchange'.[13] Though this display by Coxwell had been considered by the Committee Secretary to be something of a farce, the spectators had been kept amused and the fee for the performance was duly paid.[14]

Events at Bradford did not turn out quite so well. The craft that arrived to be flown at Peel Park was Coxwell's new creation. Stewart came by train as an escort for the balloon, and Brown promptly dubbed the craft *Napoleon*. For all its striking appearance and supposed technical advantages *Napoleon* would not get airborne when filled on the 17 and 18 June. It seemed not to have the gas-retention qualities Coxwell had claimed for it. On the Saturday after Whitsun, it was taken directly to the gasworks at the Canal. Inflated there using the more reliable equipment provided by the gasworks, the *Napoleon* rose on its maiden flight with a Mr Oldfield, Stewart and Coxwell on board. Coxwell was convinced that Bradford had some sort of jinx about it, which he referred to as 'an evil Aeronautical destiny'.[15]

Stewart's involvement in the Bradford undertaking riled Brown, and the increased commission Coxwell paid had not placated him. Stewart had displaced Brown from the car during the final, successful, ascent in *Napoleon*. Stewart had then remained in Bradford oiling the less than gas-proof *Napoleon*, a task with which Brown could well have coped. Brown was to have some small revenge for all these slights, though not through his own initiative.

Stewart suddenly received a summons to proceed with all haste to the Norwich Gardens. The Gardens had an arrangement with Coxwell to which Brown was not a party, and, through his own mismanagement Coxwell had contracted for two balloons to ascend on Monday 7 July. The only way he could cope with this *faux pas* was to allow Stewart to take up the *Napoleon*, while he flew in the *Sylph*. Brown would gladly have volunteered, but he had not yet successfully flown solo. But, as it turned out, the task was to prove too much for Stewart and he totally botched the job. During the inflation he had allowed the valve in the crown of the balloon to become twisted, and after its recent re-varnishing the balloon this time proved to be well sealed. It was incautiously allowed to stand in the sun for two and a half hours. This caused the gas to expand and the pressure proved too much of a strain for the twisted valve which burst. The balloon collapsed ingloriously. Coxwell was furious and later confided to Brown his considered opinion that Stewart's effort reflected badly not only on that man's reputation, but on his own.[16] Brown was able to offer Coxwell some consolation when he secured a booking from the Newhall Gardens, Attercliffe, for an ascent with the *Sylph* on 28 July. This time Stewart was not in attendance. With two other passengers on board, the balloon made a successful flight over half an hour in duration and covering eight miles.

All of this was but small comfort to Brown, who still felt aggrieved that Coxwell had not displaced Stewart from the *Napoleon* in his favour for the third Bradford ascent. He was also of the opinion that his skills could have been utilised more fully in the preparation for those attempts. Inexplicably, during August, Stewart was again entrusted with the *Napoleon* at the Pavilion Gardens, Woolwich. This time, apparently having learnt a little from his previous disastrous attempt, he carefully supervised the inflation of the balloon and had the craft successfully filled. With two passengers on board, Stewart took off and promptly rammed the bandstand, ripping the balloon's envelope, bringing it crashing heavily to the ground before it had even cleared the park.[17] Finally Coxwell had had enough of

Stewart's incompetence and the master/mentor relationship was ended. Coxwell feared too that, in consequence of this disaster, his own working relationship with the Woolwich pleasure gardens would also end. It was some small consolation, then, when Coxwell confirmed Brown in his position as agent, asking him to place an advertisement in the regional newspapers, in an endeavour to drum up business outside of London:

> *Balloon ascent by Mr Coxwell*
>
> Managers of public gardens, and speculators in fetes, etc. wishing to make engagements, are requested to apply for terms to Mr C.H. Brown.

Notes

1 Coxwell to Brown, 15 March 1856, *Correspondence*: 219.
2 'Leeds Royal Gardens', *Leeds Intelligencer* 29 March 1856, sec. Local and Other News: 5.
3 Since his own experience of such 'sabotage' at Bradford, Brown was always conscious of the need to keep spectators away from an inflating balloon.
4 Coxwell to Brown, 18 March 1856, *Correspondence*: 219.
5 'Local and other news', *Leeds Intelligencer* 29 March 1856: 5.
6 Coxwell to Brown, 1 April 1856, *Correspondence*: 221.
7 The Arboretum had been officially opened on 11 May 1852, but the Committee may have opted to hold the anniversary on the Whitsun. 'Leeds Datebook', 10 August 2001. <http://www.archivecdbooks.com/booke/datebook/samples>.
8 Brothers to Brown, 7 May 1856, *Correspondence*: 224.
9 Coxwell to Brown, 9 May 1856, *Correspondence*: 224.
10 There is no record of anyone other than Brown making such figures at the time – but someone was bound to have copied his idea. Brothers to Brown, 15 May 1856, *Correspondence*: 225.
11 Coxwell to Brown, 2 June 1856, *Correspondence*: 228.
12 Coxwell to Brown, 8 June 1856, *Correspondence*: 231.

13 Henry Coxwell, *My Life and Balloon Experience*, Vol.II (London: W.H. Allen & Co., 1889) 3.
14 Brothers to Brown, 22 June 1856, *Correspondence*: 232.
15 Coxwell to Brown, 1 July 1856, *Correspondence*: 233.
16 Coxwell to Brown, 17 July 1856, *Correspondence*: 235.
17 Footnote, *Correspondence*: 237.

CHAPTER 11

Steady as she Goes

Though having been confirmed in his position as Coxwell's booking agent, Brown could find little work to offer to his client. The failures at Woolwich for which Stewart was responsible had reflected on Coxwell, as he feared, and his professionalism had been brought into question.[1] On the other hand, Brown had acquired something of a reputation as a manufacturer of novelty balloons, floating giant manikins and parachutes, and this small business generated sufficient extra income to make it a worthwhile undertaking. The notoriety attached to this work was some small compensation for Brown not actually getting into the air. Coxwell took Brown's small balloons with him to use as additional attractions for the amusement of spectators and also to release as a distraction in the event of a failure of the *Sylph* or *Napoleon* to get aloft. This balloon-making, however, was not without its trials.

During September 1856 Henry Gore ordered a number of small balloons which he intended to release during the celebrations to mark the opening of a new gasworks at Worksop in south Yorkshire. The balloons were promptly manufactured and delivered, but Gore subsequently reported that they had not performed as promised. According to Gore they had not got aloft and he declined to pay for them. There the matter rested for three months. Then, on the 23 December, Brown received a letter from a gentleman at Brockley in Lincolnshire, advising him of the recovery of a small balloon from Brocklesby Park, and enclosing a card he had found attached to the balloon. This identified it as one of those sold by Brown to Gore. That very evening Brown chanced upon Gore who was back visiting Leeds. Armed with the identifying card, Brown challenged him to explain how one of the 'failed' balloons seemed to have been aloft for three months. Gore shamefacedly confessed to having used the balloon just some weeks before at another event at Paull some forty miles from Brockley Park. He promptly paid over all the money outstanding on the balloons.[2]

With Coxwell's ballooning in the doldrums he called an end to his aerial endeavours for the year. He extended an invitation to the Browns to visit him and his wife in London at Christmas. Brown, however, was unable to accept. He was pressed for funds and there were no excursions being offered by the railway at that time of year.[3]

Christmas 1857 was not a cheerful affair in the Brown household. The clerical work undertaken by Brown at Barr and Nelson's was providing a steady income, and the administrative work for Clapham at the Royal Gardens augmented that wage, but now the family was to be increased with Eliza expecting their fourth child. There was a need to supplement the budget even further. Benjamin Brown was more than ever opposed to his son's involvement in ballooning, though Charles' brother Edwin was cautiously supportive. Brown now put his mind to developing another income-generating undertaking. Over the years he had proved himself adept with the pen, and had also perfected his skill as an artist-engraver. He considered that a venture into the realms of publishing might prove profitable; it was an area in which he had previously dabbled. However, before he could advance this idea to any degree, he was caught up with Coxwell's request for provincial engagements. Would Clapham take him on again in the new year? And what of the proprietors of the parks from which successful ascents had been made the previous year?[4] Brown quickly obtained an engagement for Coxwell to make a night ascent for the Royal Gardens, but to his chagrin Coxwell turned this down as not being worth the risk involved. After some consideration, however, Coxwell realised this was not a very sensible response. He confided to Brown that 'on second thought, it would not be wise or right to oppose Clapham, unless engaged by some one else'.[5] As Coxwell vacillated, Brown took the opportunity to put forward his publishing proposal.

A number of issues of *The Balloon or Aerostatic Magazine* had been put out by Coxwell during 1845, and Brown had been an enthusiastic reader and would-be contributor to the little publication. He now suggested that the magazine be resuscitated, and he would undertake the printing and engraving. A startled Coxwell replied:

> I was rather surprised to hear of your determination to 'start printer' – never having heard you express any bias for the handling of lead, either in the shape of bullets or devil's type. I thought the engraving process was the one selected to engage your spare moments. As to the *Aerostatic Magazine*, I confess a sort of pride would lead me to delight in its continuation, whilst, however, common sense and past experience unite to assure me that each attempt would be money out of pocket ... I don't know whether you read in last week's *Punch* a paragraph headed 'How to remove ink stains – Publishing on one's own account' is certainly the way to check this propensity.[6]

This was not the response Brown had anticipated, and within a couple of weeks he had had a falling-out with Coxwell, perhaps to some degree in consequence of that man's failure to warm to the publishing proposition. The row concerned the production and sale of Brown's small balloons. Coxwell had innocently asked Brown to provide a number of balloons for an engagement at Derby that he had negotiated on his own behalf. Coxwell told him he would recompense him directly, as the promoter preferred to make an all-up payment. Brown construed this as a move by Coxwell to take control of his pilot balloon market. The more commercially experienced Coxwell good-naturedly attempted to counter the other's prickliness by stating he was only passing on an order for Brown's benefit, and asked that the whole silly affair be let drop.[7] Brown calmed himself and made no further comment, and even provided further balloons to Coxwell as well as miniature parachutes and inflatable manikins, but he was irked that Coxwell was now to all intents and purposes acting as his own agent, leaving him to deal with the few engagements on offer at Leeds or Bradford.

In an effort to resolve what he saw as differences between himself and Coxwell, Brown decided to go to London and have the matter out. Far from settling the affair the move made the whole situation worse. Mrs Coxwell invited him to stay at their house, but somehow Brown imagined he was unwelcome. When Coxwell discovered the misunderstanding, he immediately wrote restating his offer of accommodation in the clearest terms, and offering to pay Brown's rail fare down to London. Pacified, Brown took up the offer, though as it transpired, this did not really help matters. On the day he chose to make his visit, Coxwell had been engaged to fly from Charlton Hall, the residence of Sir Thomas Wilson in Greenwich. Coxwell took Brown along with him to witness the ascent though all the places in the balloon's car had been reserved for Sir Thomas' guests. Intent on making a good impression and having a successful ascent, Coxwell then proceeded to pay more attention to the balloon than he did to his guest. Brown left in a huff and went back to Leeds. Coxwell quickly wrote to apologise, explaining he had been 'anxious and fidgety' because, at the time, he believed he had insufficient ballast for the flight. Again Brown regained his equilibrium, though the constant bickering placed an increasing strain on the relationship. Both parties tried to set themselves back on a normal business and social footing.

Brown was still keen to draw Coxwell into a publishing venture. This time he proposed that rather than attempt to revive *The Balloon* he would produce an entirely new publication in the form of a broadsheet of aerostatic news. Coxwell allowed himself to enthuse about the publishing idea and he threw in the suggestion that an almanac format would work even better: a yearly publication with aerostatic news, and details of remarkable balloon trips, together with advertisements for the principal pleasure gardens. He went as far as to offer Brown the use of his old engravings from *The Balloon*.[8] A few weeks later there appeared the first, and only, edition of *The Aerostatic Remembrancer*. It contained the balloon news from the previous months and a few advertisements. As proposed, Brown had set it up and printed it, but it served mainly as an advertisement for himself and Coxwell.[9]

Once again events took an irritating turn. However, on this occasion it had nothing to do with the relationship between the two men. The use of trapeze artists to add lustre to balloon ascents had been a feature of 'theatrical' ballooning for many years. It was a performance intensely disliked by Coxwell, though he had at times buckled under pressure from garden proprietors, and taken up such acrobats. Both men held the view that it discredited serious aerostatic endeavour. Coxwell believed that Simpson at the Cremorne Gardens, where he was increasingly being engaged, was about to re-introduce the balloon/trapeze ascents as an attraction. He drafted a letter and suggested to Brown that he could edit it and send it to the *Morning Advertiser* in London with copies to the provincial papers. It called attention to the revival of 'acrobatic performances under balloons', which 'disgraces those interesting machines'. Rather pompously, Coxwell wrote: 'The voyages of Mr Charles Green and of Mr Henry Coxwell somehow or another shed lustre on aerostatics, but the freaks of mountebanks throw discredit on ballooning, and lower proprietors of gardens in public estimation'. Although the letter was to be sent anonymously, this clearly was an instance where Coxwell's indignation had gotten the better of his literary judgement. Wisely Brown ignored Coxwell's draft and composed his own missive for the newspapers. The finished work appeared in *The Era*, the London theatrical newspaper, on 26 July 1857:

> Sir, — I am sorry to find that the trapeze performance under balloons is again revived, both in London and Sheffield, notwithstanding the orders given three years ago to the police by the Secretary of State, to suppress it. Setting aside the foolhardiness and uselessness of the thing, is it right that one place of amusement should be allowed to produce the exhibition and another not? Mr Wardell closed Vauxhall on account of the prohibition of trapeze ballooning, he being unable in consequence of that prohibition, to produce attractions offered by other establishments.
>
> Yours obediently, Aero.
> July 23, 1857

The editor rather primly noted that, although such an event had been advertised, the 'disgraceful feat' did not take place, the proprietor claiming the weather was too severe, and 'since that time there had been no trapeze ballooning in London or its vicinity'.[10] That matter resolved to their satisfaction, Brown continued to manufacture his balloons and Coxwell continued to make ascents in London and, when Brown could arrange it, at venues adjacent to Leeds and Bradford.

Notes

1 Coxwell to Brown, 3 September 1856, *Correspondence*: 238.
2 Footnote, *Correspondence*: 245.
3 The normal Leeds–London single fare in 1856 was 30/- single, excursions would be run with fares as low as 5/- return. 'Rail History' 12 October 2001.<www.mikes.railhistory.railfan.net>.

4 Coxwell to Brown, 11 February 1857, *Correspondence*: 246.
5 Coxwell to Brown, 10 March 1857, *Correspondence*: 248.
6 Coxwell to Brown, 20 April 1857, *Correspondence*: 251.
7 Coxwell to Brown, 22 May 1857, *Correspondence*: 252.
8 Coxwell to Brown, 25 June 1857, *Correspondence*: 255.
9 Footnote, *Correspondence*: 257.
10 'Editorial Notice', *The Era* 26 July 1857.

CHAPTER 12

The Proposition

In early September 1857, disaster struck Coxwell. He had hired out his balloon, the *Sylph*, to Simpson at Cremorne Gardens, sacrificing fares for a set price. Intent on getting his money's worth, Simpson overloaded the car with passengers. The *Sylph*, failing to gain sufficient altitude, flew into a weathercock and the envelope was torn. Fortunately the craft made a slow descent, and this gave the whole affair the appearance of being arranged and well managed, but the balloon needed to be laid up for repair. It was taken to an old sail loft that Coxwell had hired as a storage place for his machines. One night, while the balloon was still being refurbished, a neighbouring ship-builder's shed caught alight, and the fire spread to the sail loft and destroyed the *Sylph*. Coxwell, uninsured, was left with only one balloon, the original *Napoleon* (by now renamed *Tavistock*) which was in such poor condition as to be unreliable.[1] This misfortune provided Mrs Coxwell with a long awaited opportunity to have her say, as she was firmly set against her husband continuing as an aeronaut. This was the very time when Coxwell was sought out by George Coppin, the Australian entrepreneur. 'I have just had an offer from Mr Coppin', he informed Brown, 'to go with him to Australia, for one, two or three years: I have declined – not having proper balloons – and for domestic reasons beside. I should like the trip immensely, but don't see my way clear'.[2] Coppin's presence in London had not gone unnoticed by Brown.

Coppin was nonplussed by Coxwell's refusal to consider travelling to Australia. He had after all made a very generous offer and 'would not stand particular for a few hundred, more or less'. But Mrs Coxwell was firmly set against her husband leaving England – she was not of a strong constitution and Australia was a foreign, inhospitable place – and that was that. In the face of this setback, Coppin still needed to purchase two balloons. He had, after all, promised his patrons back in Melbourne that there would be aerial ascents in the new year from his pleasure gardens at Cremorne. So, as an alternative to having him as an aeronaut, he sought to commission Coxwell to make the balloons for him. Again Mrs Coxwell stepped in and put down her foot. She was quite sure Mr Coppin had no idea of the mess and confusion associated with the construction of a balloon building. And her husband certainly was not going to get involved in the manufacture of *two* balloons.[3]

But with the prospect of a deal of money being made out of the affair, Coxwell was not prepared to pass up this opportunity so easily, whether his wife approved or not. Brown had gotten wind of Coppin's need for two balloons and, without any reference to Coxwell in the matter, had written quoting a figure of £160 for a muslin balloon which, as it transpired, matched exactly the price Coxwell had initially proposed to Coppin – that had been before Mrs Coxwell had intervened and put a stop to the business. When Coxwell got to hear of Brown's involvement he quickly contacted him, and suggested a partnership. The idea was that Brown would fabricate the balloon from a pattern that Coxwell would make. Coxwell would also arrange for the netting and the valve to be made in London. The injured Chambers, now looking to return to aerostation, could be relied on to produce netting of good standard. After expenses Brown and Coxwell would then share the balance of the £160 pro rata. All of this could be done out of sight of the antagonistic Mrs Coxwell. Since there was a little over a month before Coppin's planned departure on 15 October, they had to move quickly.

What Coxwell had missed in the excitement and bustle of all of the to-ing and fro-ing was the fact that Brown wanted to undertake the whole commission himself: the manufacture of the balloon and the journey to Australia to fly the craft. When he discovered this, Coxwell was perplexed. Previously, when the two men had discussed future prospects, Brown claimed a career as an aeronaut was no longer his prime objective.[4] He had a pregnant wife, with the infant due early in the new year. Family must have priority.

Things were now, however, appearing in a different light. Given the prospect of taking a balloon to Australia, and becoming a pioneer aeronaut, he confessed his enthusiasm had been fired again. He was gripped by the aeronautic fever, this time more strongly than ever. Coxwell, having been firmly reminded of his own marital responsibilities, undertook to give Brown some fatherly, and prophetic, advice:

> I am quite certain by my own experience that the only way to cure the mania is to allow it full scope, and then it appears in its reality something so different to what imagination or ambition pictured it, that a useful lesson is to be learnt – a lesson which it is useless to inculcate until a fair and dispassionate comparison of risks and rewards enables the party learning to judge which is wisest and most profitable.[5]

How he imagined Brown could undertake such a fair and dispassionate comparison before actually committing himself to the Australian undertaking was not clear. Having said his piece, however, he undertook to recommend Brown to Coppin, both as manufacturer of the balloon and as the ideal aeronaut to fly the craft in Australia. Accordingly he wrote to Coppin:

> Since I wrote to you I have heard from my friend, Mr C.H. Brown of Leeds, and I find he is willing to undertake the whole affair. Of this I was not aware when you mentioned his name to me – but all I can say is, you won't find anybody upon whom you can place so much reliance and I am certain we have no Aeronaut in the present day who has made Aeronautics so much his study. His experience with me has not been extensive, but he had had a balloon of his own, and has associated with other Aeronauts. Of his trips with me I can speak highly, and I have no doubt he will prove a credit to you should you engage him.[6]

George Coppin was an Englishman beset by a keenness never to miss the main chance. He had left home at the age of twenty-three and sailed to Australia to follow his craft as a stage comedian. Reaching Sydney in March 1843, he quickly built a reputation for himself on the local stage, and made a small fortune. He then moved to Hobart Town, and in time followed the gold rush to Victoria. He experienced mixed fortunes over the next years, his financial successes being matched by an equal number of failures. Insolvent in 1851 he finally settled with his creditors and sailed for England.[7] By December 1854 he was back in Melbourne. He entered a partnership with the eccentric Irish actor, Gustavus Brooke, and the pair had purchased the Theatre Royal and the Olympic Theatre in Melbourne.[8]

The Cremorne Gardens were Coppin's independent venture, however, and he was singularly set on making an immediate success of that undertaking. The Gardens, laid out over a ten acre plot, stretched along the north bank of the Yarra River. They were touted as 'the Playground of Melbourne', and modelled on the London Cremorne Gardens.[9] Coppin had set himself to equal that establishment, by means of a menagerie of Australian animals, an artificial lake, a theatre for concerts and vaudeville and a large 'Parisian' dancing platform.[10]

Coppin had engaged John W. Anson, an actor and an adept with finances, to be his agent in London.[11] Coppin was prepared to take up with Brown on Coxwell's say-so. Coached by Coxwell, Brown requested a monetary advance against the purchase of materials. Anson sent off a postal order for £50 to cover initial expenses, and at the same time warned that everything must now be ready by 4 October. An agreement of service had been drawn up, and Brown was invited to travel to London to sign it.[12] He quickly went down and meeting Coppin on 24 September, signed a contract to fly balloons for him, and to be based at the Cremorne Gardens in Melbourne.[13]

With that business out of the way, no time was wasted and passage for Coppin, Brown and the balloons was booked on the *Overland*, the departure date being settled for 15 October. Coppin wanted to return to Australia as quickly as possible. He was intent on keeping the promise he had given the public of Melbourne that he would greet the New Year with an outstanding novelty – Australia's first balloon ascent; and he hoped that by taking passage on the *Overland* he would get back to Australia in time to fulfil his undertaking. The route was by sea and land from Portsmouth to Suez, there to pick up a ship to Australia. This route was calculated to save up to five weeks on the time taken by following the normal sea route via Cape Town.

Following his trip to London to sign the contract, Brown discovered that he was not to be the sole aeronaut employed by Coppin. Joseph Dean, a former wire maker turned journeyman aeronaut, had also signed on for the Australian trip. Dean had flown on many occasions, at times using the alias 'Captain Bedey'. He was a hard-drinking man, who had a habit of sniffing coal gas during balloon inflations, and this habit and his drinking had left him with something of an unsteady reputation. Coxwell was uneasy about Coppin's engagement of Dean. He had worked with Dean on occasions and was upset by his lack of professionalism.[14] Brown also was convinced that the man was given to sharp practice, puffery and self-aggrandisement. Brown's sense of caution was not misplaced, for in the course of time Dean was to occasion him a considerable amount of heartache. But all that was in the future and there was little time to brood over such things in the rush to prepare for the journey to Australia.

Brown was to take charge of the balloons and travel with Coppin, while the pregnant Eliza and the three children would make the whole trip by sea on board the clipper, *Swiftsure*, due to depart on 15 October from Plymouth. Dean would travel on the American ship the *Shakespeare*, which would sail from Liverpool on 22 September, putting in at Plymouth on 27 September.[15] Other artists engaged by Coppin were to travel out independently. When in London to sign his contract, Brown met one of these artists, John Anderson, *The Great Wizard of the North*. 'Professor' Anderson had performed his Magic Bottle act for Her Majesty at Buckingham Palace, producing from the bottle any drink requested by members of the audience.[16] Off the stage he had a dourness of character which, allied to an affected air of superiority, left Brown feeling that he did not much care for the man.

The travel arrangements finalised, Brown could not contain his excitement and promptly wrote to his regular correspondents, MacSweeny and Wise, advising them of his change of fortune and his impending departure for Australia. MacSweeny replied that he 'was pleased to find that you have got an engagement that will tend to your interest, and to the interest of aerostation. I am sure', he added, 'your zeal for the science will lead you to adopt every improvement'.[17] The postal delay between England and the United States precluded Wise from sending any laudatory missive that would have caught Brown before his departure for Australia.

But all had not progressed as smoothly as could have been hoped. Coppin had originally intended to purchase two new balloons, but it was becoming clear that the balloons he had ordered could not be constructed in time to meet the departure date. Brown and Coxwell then struck on the idea of giving the *Tavistock* a hasty re-furbishing and offering it to Coppin. Intent on obtaining two balloons, and having no experience in matters aeronautical, he was prepared to accept the second-hand balloon on trust.[18] Coxwell arranged for the machine to be delivered to Anson at Bow Street. Brown knew that Coppin was being sold a defective and potentially dangerous craft, but the excitement of the whole Australian venture helped still his doubts over the matter.[19] As a further balm to his conscience, he imagined that there would be time to make the necessary overhaul of the craft when he arrived in Melbourne.

At Plymouth, on Thursday, 15 October 1857, Coppin and Brown set off across the English Channel, bound for Le Havre and onward to Marseilles. The plan was then to go on via Malta to Cairo. They were to board a P & O steamship at Suez. Brown's adventures in the Antipodean world were about to begin.

Notes

1 Coxwell to Brown, 8 September 1857, *Correspondence*: 264. The *Napoleon* had been flown from Woolwich to Tavistock in July 1857, a distance of 200 miles. The flight had taken five hours. Immediately afterwards the craft was renamed.

2 Coxwell to Brown, 10 September 1857, *Correspondence*: 265.

3 Coxwell to Brown, 10 September 1857, *Correspondence*: 266.

4 At the time Coxwell recalled – 'I thought from what you said to Mrs Coxwell and to me that you had abandoned any idea of [becoming an aeronaut]' Coxwell to Brown, 12 September 1857, *Correspondence*: 265.

5 Coxwell to Brown, 14 September 1857, *Correspondence*: 267.

6 *Correspondence* 267.

7 'George Seth Coppin.' *Australian Dictionary of Biography* (Melbourne: Melbourne University Press, 1966) 460.

8 *Australian Dictionary of Biography* 459.

9 Agnes Paton Bell, *Melbourne – John Batman's Village* (Melbourne: Cassell, 1965) 83.

10 William Lawrence, *The Life of Gustavus Vaughan Brooke* (Belfast: W & G Baird, 1892) 178.

11 The Adelphi Theatre, 20 April 2001. <www.emich.edu/public/ /adelphi_calendar1865>.

12 Anson to Brown, 21 September 1857, *Correspondence*: 268.

13 *The Coppin Papers*, Mss 8827, Box 12/18 La Trobe Australian Mss. Collection, State Library of Victoria, Melbourne.

14 Coxwell to Brown, 8 September 1857, *Correspondence*: 263.

15 *Shipping List*, VPRS: 108558: 2.

16 'Professor Anderson in Boston', *Gleason's Pictorial Drawing Room Companion*, 13 September 1852: 193. The command performance was given during 1843.

17 MacSweeny to Brown, 26 September 1857, *Correspondence*: 271.

18 Coxwell to Brown, 30 September 1857, *Correspondence*: 278.

19 Coxwell told Brown that the *Tavistock* would likely be of no use for the 1858 season, and was certainly not reliable. Coxwell to Brown, 8 September 1857, *Correspondence*: 264.

In London

The place was cheery and noisy. I could just hear the orchestra over the clatter of glasses and the laughter. I was standing at the bar in the back of the stalls. The Winchester had always been one of my favourite music halls. It was a very well turned out place, with bars and such like, but that's not the reason I took to it. No, I was particularly fond of the Winchester because Mister Preece, the proprietor, had given me my first opportunity to appear there as a performer. That was twelve months previous. Now I had been promised that, when the current bill was finished in a week's time, I would make another appearance. I intended to give a parody on Mister Foster's song, 'Old Dog Tray.' You know, the one that goes: Old dog Tray! always hungry!

I had taken a chance in turning to the stage. My father had intended me to follow him to the bench as a carpenter, but I wanted to sing. Not just for my own pleasure, but I had this desire to perform. Perhaps it's something I inherited from my mother. She was always to the fore in any family frolic. As my father would say to her, 'You've got no shame about you, lass!' Truth to tell, up to now I'd not made a great deal of money from my venture into singing, for I hadn't had that many engagements, so I was glad for this opportunity to get back on the boards. London can be a lonely place when you aren't native to the city. So, here I was, back at the Winchester getting the feel of the place again. The acts I'd seen weren't all that flash and I knew I could hold my own against them. The Great Wizard of the North was to have been on the same bill as me the next week, but he'd pulled out. Now, his would have been a hard act to follow, 'cos he can knock the stubholders bowlegged, especially if he'd done his Great Gun Trick – the one where he catches a bullet fired at him from a musket. But there was nothing of that quality playing this night, and the audience were sitting on their hands and getting on with the drinking. And then – you know how it is when you can pick up something over a hubbub – well, I heard this Yorkshire accent. 'One of my own,' I thought to myself, and had a look along the bar. Straight away I spotted Johnny Anson from the Adelphi. I'd met him a few times and he thought he could get me work there. It's not a music hall, but they run some stuff that uses singers. They'd just done a nautical play that needed some shanties sung, but I didn't get that gig. Any road, I went down to say 'hello' to him, and tell him of my good fortune in getting a run from Mister Preece, and he introduced me to this bloke. Charles Henry Brown. It sounded rather grand, but he was a nice bloke. He was the one with the Yorkshire accent. Not the 'aye oop' sort, but a kind of drawl. Sounded just like my father.

'You're on the stage, then?' I asked him.

'No, I'm an aeronaut,' he told me.

'What's that, when it's at home?'

'I fly balloons,' he said, and handed me a card with his name on it, and on the bottom it was engraved, Melbourne, Victoria, and there was a picture of a balloon. 'I made that myself.'

'The card or the balloon?' I said, by way of a joke.

'Both.' He smiled. Not a big man, though tall enough, with a neat tash. Only a few years older than me, I reckoned, even though he did sound like my father. Mister Anson had gone into the stalls to watch a troupe of Japanese acrobats, so I was left with Mister Brown.

'You're from Yorkshire – not from the colonies,' I said.

'True. But I am about to depart for Australia. I have been engaged by Mister George Coppin to fly a balloon for him – in Melbourne. I had some final business to attend to with Mister Anson. My ship sails from Portsmouth within a few days. We then travel overland to Suez.' Well, I told him that would be very exciting, I was sure. Then I couldn't resist telling him that I was in the entertainment business myself. That got him rather upset.

'You must understand,' he said, sounding a little uppity, 'Aerostation is not concerned with providing amusement for the public, but is a serious scientific undertaking. The aspect of entertainment and spectacle is, in some instances, an inescapable adjunct to the undertaking.'

That sounded great – that's how I came to remember it – but I didn't know what he was on about, so I let it pass. Then he asked me how I knew he was from Yorkshire. I told him I'd spotted the accent.

'I'm from Yorkshire myself,' I explained. 'Born in Redcar, do you know it? Mother and father are from there too, though they don't live there any longer.'

'Redcar by the sea,' he smiled. 'I know it, but have never made an ascent there. The sea can prove a hazard for aeronauts. The public imagine that it is safer to descend on water than on land, but they are wrong on that account. More aeronauts have been drowned than have met with any other kind of mishap.'

From what I'd heard about balloons and such I doubted that, but held my whist. 'Here, what am I thinking about – you'll have a beer?'

'No thank you, I avoid stimulants when I can. I've seen too many in my field who've done themselves harm by indulging in intoxicants.'

I had to admit that Mister Charles Henry Brown had a way with words. I'd rather taken to him. 'I know a man who does ballooning.'

'Aeronaut,' he corrected me.

'I don't know if he is still doing it – aeronauting. That was in Dublin.'

'Dublin?' he said, surprised.

'Oh, I didn't tell you. That's where my family is living now. My father is a shop-fitter. He has a little factory down on Wood Quay.' He seemed surprised that a Yorkshire family should choose to live in Dublin. It were a sight closer than Melbourne, but there's no accounting for taste.

'I have an acquaintance who practices aerostation in Ireland. Mainly in Dublin, at the Rotunda and Portobello. I am afraid we are not presently on good terms. Mr Hampton.'

'Johnny Hampton? Lived in Jervis Street, just up from the Quay?'

'That would be my acquaintance,' he admitted. 'How extraordinary. I do not know that I would call him Johnny.'

'Small world,' I said. 'My father made frames and rollers for Johnny's … Mister Hampton's panoramas. I say, why don't you go to Dublin and fly? Easier to get to than Australia.' That was another little joke, but Mister Brown appeared to take it seriously.

'I had at one stage anticipated I would do so, but experience has now convinced me that Mister Hampton would strenuously oppose such a move.'

I didn't want to pry, and I reckoned I'd been a little too personal already. I decided to change the subject. 'How did you meet Mister Coppin? I've heard of him, but I thought he was in theatres and not airships.'

'It was by chance that I got to hear of his search for a manufacturer of balloons. And then a good friend, Mister Coxwell, put me in touch, and matters developed from there.'

'Why, I've seen Mister Coxwell go up in his balloon from the Surrey Gardens, just down the road from here. He did signalling with flags and let out pigeons. It was quite a sight.'

'It was a scientific experiment in aerial communications for use in modern warfare.' There he went again, I thought to myself. He has a lovely way with words.

'So, it's Australia then?' I said, changing the subject.

'Yes, Australia. There is the challenge of being the first to make a successful aerial ascension there. And then there is the vast interior to be explored, an undertaking which I can only see as successfully achieved through the use of balloons.'

The band then started to play the introduction to 'Old Dog Tray.' Then on stage bounced Teddy Wright, from the Adelphi. He was moonlighting, and, bugger me, he'd stolen my song – he was singing the parody and not Mister Foster's serious version. I was mad. I'd have to find another number for next week. When I recovered from the shock I saw that Mister Brown had tipped his hat and was moving off. 'Remember me to Mister Coppin,' I called after him. 'Maybe he can find me a gig in Melbourne – and good luck with the balloon.'

'It will be skill and application that takes me aloft,' he said and walked away. I loved the way he used words. I knew there was something in French I should say. 'Bon', something or other, but I'd forgotten it. I'd also neglected to ask him if he'd a family that was going with him to Melbourne. I'm sure he would have mentioned it if he'd a wife and youngsters.

CHAPTER 13

The Overland Route

Marseilles. Sunday, 18 October 1857. I am putting pen to paper to write in a style that is unfamiliar to me, as it is my intention to keep a diary of the journey from Marseilles to Melbourne.[1] This being my first time abroad, I may claim it to be the journey of a lifetime, a claim which I previously made for my maiden voyage aloft. We are embarked on the RMS *Cambria*. It is a steamer of the Cunard fleet, but it has been chartered by the European and Australian Royal Mail Company. I am told we can travel at 9 knots, and if we can sustain that speed we will be at Malta post-haste. As I stood on the deck this evening after most of the passengers had gone below, I could not help reflecting on the good fortune which had secured for me this opportunity. The pity is that I am not able to undertake this adventure with my dear Eliza and the children. She has sailed on the *Swiftsure* from Plymouth. The ship was in port when I took Eliza there. It is a handsome steamer and will be on the high sea even as I write as it was scheduled to sail on the fifteenth of this month, the same day as we left Portsmouth, but Mr Coppin declined to take passage on it. He insisted that I travel with him on what he calls 'this speediest of routes', which will take us overland from Alexandria to Suez. The purpose of our haste lies in the fact that Mr Coppin gave to his patrons in Australia a personal assurance that he would bring with him a brilliant cavalcade of new talents in time for the Melbourne Christmas theatre season.[2] And it seems he made that promise in such particulars as have already provided him with his share of heartbreak. For he promised to obtain the services of Miss Cushman, the American actress, who made something of a name for herself by playing Romeo in Edinburgh and also Madame Celeste, the *danseuse*, whom I once saw in London, and Mr Charles Mathews the actor.[3] But all had departed to tour America before Mr Coppin could engage them. So there is myself and

Mr Anderson, the famous Wizard of the North, secured to Mr Coppin by contract, and he intent not to let us out of his sight, or so it would seem.

The journey through France was uneventful, though far from pleasant. There were a myriad of officials at the custom house in Le Havre who, when not intent on seeking additional recompense for their services, were overcome by an unhelpful curiosity as to the contents of the two wicker cars, packed with the *Australasian* and the *Tavistock* along with all the necessary paraphernalia.

The post-coach – or as the French would have it, the Diligence – in which we journeyed provided as comfortable and speedy a vehicle as our own Parliamentarys, and the roads are uncommonly good. But while interesting, the travel is troublesome and not necessarily reliable. I saw no less than three coaches turned turtle in the space of a few miles as we travelled between Paris and Macon, though for these disasters I would be inclined to blame the drivers rather than attribute fault to the vehicles or the roads.[4] But, that aside, I am now safely ensconced on board the steamer. Mr Coppin has already spoken to the Captain, complaining that the ship's accommodation is crowded. I gather Mr Coppin fears that the many passengers bound for India are threatening to displace him from his cabin.[5] He has argued that he and I must be accommodated in the berths we have already reserved, as our journey is even longer than that to Calcutta.

Mr Coppin has decided we travel saloon, as there is then less of a problem to be experienced with the great weight of baggage that we are bringing with us. I have with me in my cabin all my letters, which are packed in the watertight manner Mr Hampton devised (at least that is one good thing which came out of my dealings with that man.)

Monday, 19 October. The movement of the ship is now, for some reason, so very different from that which I experienced on the short passage from Plymouth to Le Havre, and it has left me feeling unwell. There are smells from the kitchen (or should I call it the galley?) to which in my present state I am singularly sensitive, and with which I could well do without; cabbage boiling, mutton roasting, and, unless I am seriously mistaken, there is also a curry being brewed.[6] How unlike the smooth and languid passage of a balloon. Perhaps, when the problem of navigability has been overcome, a journey such as we now undertake between France and Egypt, will be managed in a balloon, the size even of my own *Celeste*.[7] I was even for a moment at Marseille tempted to joke with Mr Coppin about such a prospect, but he did not seem to be in such a humour as to be appreciative of my jest, so I let it pass.

Thursday, 22 October. We are to dock at Malta tomorrow, and there will be time in which to visit whatever sight there may be to see on the island. I will keep my own company should I undertake such a trip. Mr Coppin has his own friends on board. I cannot say I care much for Mr Anderson. He is a Scotchman and has a sharpness of speech and that dourness of character which is a mark of his race. That he has performed for Her Majesty at Her command has, I fear, encouraged him to adopt a superior manner which ill becomes him.[8]

Friday, 23 October. We are in the harbour of Valletta. I have fully recovered from my seasickness, having become accustomed to the peculiar motion of the ship. I have returned from a walk about the town. It is not large, though there are some interesting forts that were built by the Knights Templar. The French have left no mark of the short time they occupied the island. This island has now been in our possession for over fifty years. There are many English soldiers and marines in quarters. The heat is oppressive, and I am anxious as to how it will affect the balloons. Mr Coxwell insisted that they be varnished before the trip. I argued that this was not a good idea, as the varnish, no matter what preparation is used, does not stand up well to heat. And particularly when the balloon is compressed, there is the potential for cracks and tears to occur. Coxwell countered that a new coat of varnish would be necessary to spruce up the *Tavistock*, giving it an unused look. It is an old balloon and this Mr Coppin knows, but Coxwell was anxious that it not appear as being in such poor condition as it really is. But this is something he must deal with in his own conscience, as he is the one who sold that balloon to Mr Coppin. For my part I can most wholeheartedly vouch for the quality and durability of the *Australasian*.

There are some passengers who approach the crew to obtain details of our latitude and longitude. This is an undertaking that holds no attraction at the present, since we are for the most part never long from the sight of land. Perhaps when we reach the expanse of the Indian Ocean, it may prove an interesting exercise.

Sunday, 25 October. As I write we are standing off from the port of Alexandria. There is the utmost bustle and confusion on deck as the crew prepares to dock, and I am best below decks for the moment. It is here that all the passengers, baggage, cargo and mail are to be landed, and we then go overland to Suez. I am remiss in not having before mentioned that we are to take passage on the RMS *Emeu* from Suez.

What a mix Alexandria is. There are people from all nations, and the town itself shows what I imagine to be the architecture of all countries. There is a French quarter here where the houses are the very twins of those I saw in my short time in Marseilles. But the curiosities are few and easily seen. The hotels are all whitewashed with trellised windows to counter the heat.[9] I am surprised at the number of Negroes that I have seen here, quite as many as there are Arabs. When all the cargo is finally unloaded we are to set out for Cairo. I sought to supervise the handling of the balloons, but the dockers, brown faced, half naked and sulky, would take instructions from no one but their overseer, and he, a long robed, turbaned and large Egyptian, appeared to have no knowledge of English – or at least feigned such ignorance.

I am discovering that in consequence of the nature of our journey to date, I neglect making regular diary entries. The passage thus far has made a surprising call on one physically, and I am suffering a little soreness in my arms to which I would not usually admit. Now that our baggage and the several items of cargo have been released from the realm of the dockers, I am obliged to superintend its handling, for fear of it being seriously mismanaged by the troop of miscreant Arabs. This motley band regularly turn toward Mecca and fall on their faces, when they should properly be attending to the baggage and the barges on which we will travel the Mahmoudia Canal to Cairo.

Monday, 26 October. I was mistaken about the leg of the trip we undertook on the Mahmoudia Canal. The barges, or more properly fly-boats, were hauled by a small steamer up to Atfeh, where we now are.[10] It is a little village of mud cabins, and a few small bazaars. I had anticipated that we would be taken through to Cairo, but it now appears we must again transfer ourselves and our impedimenta to another vessel, as we have reached the lock that forms the junction of this canal with the Nile. We are now to board a steamer that will take us on the river through to Cairo.

Tuesday, 27 October. It has taken us twenty-four hours to get to Boulak, which is the port at Cairo. There is a small fleet at the docks, loading and unloading with a deal of noise and bustle. And matching the minaret of the mosque there are tall chimneys from factories and foundries.[11] There is a raised road that will lead us into the city. We are to stay at Cairo and await the semaphore message across the desert from Suez, confirming the arrival at the port there of the ship bound for Australia. It is 84 miles from Cairo to Suez, and the system of signalling reminds me of Mr Coxwell and his War Balloon, though he never knew the semaphoric system and used his own special code. We are to be conveyed to the next post in covered cars. Mr Coppin has been off ahead into Cairo to affect some business. He has not spoken much to me since we arrived in Egypt, and Mr Anderson has spoken not at all. I have been left to my own devices in the management of the balloons, which I find satisfactory.

Wednesday, 28 October. This morning we were introduced to the new railway, which we were informed would take us to within seventeen miles of Suez. At the end of the line we were packed into small green omnibuses, pulled by four horses. Six passengers in each. Our baggage was strapped to camels, and the cargo piled on carts pulled by mules. It was a most uncomfortable, jolting journey. I have been spoken to by a Captain Henderson, who has travelled to Melbourne on this route before. He told me that prior to the railway opening, and that only recently, it was necessary to travel all along the desert road from Cairo to Suez, and there were seven station houses along the way. The railway now ceased at the largest of these. I was able to view it, a modest building with bed-chambers and a kitchen, but once the camels were loaded we had to press on to Suez.[12]

Thursday, 29 October. Suez is an unpleasant and dirty post. The hotel in which we are staying is kept by a German. It is a large cloistered building, with bedrooms opening onto the courtyard. Most disconcertingly, the *Emeu* is not in port, and no one seems to have any news of it. In an attempt to thaw the frostiness of Professor Anderson, I politely asked him if he had any idea what was amiss. He put his finger to his nose and said in a most unpleasant tone, 'Beware the Arnanoots!' or something that sounded very much like that. His accent is a strangle jumble of Cockney and Glaswegian. I shall not attempt to enter conversation with him again.[13]

Friday, 30 October. Another day at Suez. Mr Coppin had gone to the docks with the German hotel owner, and has just returned in an agitated state. It appears that the RMS *Emeu* is running late up the coast. It will not be here till Monday, the first of the month. I asked the hotel owner if there was anything I could do to help. He replied to me in German. I have no idea what he said. I believe he was drunk.

Tuesday, 3 November. News has reached us that the RMS *Emeu* has gone aground on a reef in the Red Sea, at a place shown on the maps as Dubberdubb.[14] We are told extensive damage has occurred. It would seem also that some of the Royal Mail has been misplaced – which does not surprise me in light of how the fellahs dealt with it while loading at the Mahmoudia Canal. There is not much by way of attractive scenery around Suez. I had heard it said Napoleon proposed a canal between the Mediterranean and the Red Sea. Such a traffic-way would alleviate the need for such a tortuous overland expedition.

Thursday, 5 November. There may be some relief in sight for our distress. The RMS *Simla* is due into Suez on the twenty-seventh of this month, and we will be able to take passage on that vessel.

Friday, 6 November. The damaged *Emeu* made it into port this morning. She is sitting very low in the water, and I have been told that she cannot be repaired here, but must go to Bombay. I would doubt that she could make such a trip. The mail, missing from Alexandria, has arrived.

Friday, 27 November. The RMS *Simla* has arrived and is discharging her cargo and the mail. As the date shows, we have been detained here for four weeks. Mr Coppin has had to send to Cairo for additional funds to sustain us during the delay. The shipping agent has done the best he can to calm the passengers, but there are one hundred and twenty very distraught and angry people sitting about in this mosquito-riddled place. I have curtains that I tuck about my bed most carefully, but these tormenting insects find their way through, I know not how. It will be a blessed relief to be upon the sea again, even if I should be as discomforted as I was previously on the trip directly out of Marseilles. Mr Coppin has been to see the Captain,

Mr Cooper, and has confirmed that we will be accommodated in the saloon, and further, the balloons can be safely stowed on board.

Thursday, 29 November. Captain Cooper has himself come with the news that we will sail on the tide today. Mr Coppin has made notes of all the particulars of the situation and says he will have it out when we finally arrive at Melbourne, and that he will raise 'a stink' in the newspapers. There is a political complexion to this, he states, and he is of the opinion that the Egyptian Transit Company may be in some ways involved.[15] I cannot see how they would be to blame for the *Emeu* running onto the reef, but I do not understand any of this. I will now go and ensure that all is in order for the balloons finally to be loaded.

Later. I have just returned from seeing the balloons stowed. The *Simla* is a handsomer ship than the *Emeu* – though I am conscious that the *Emeu* was damaged when I saw her. The *Simla* is on charter to the European and Australian Company, though it is a Peninsular and Oriental vessel, and Captain Cooper is one of their officers. She is built of iron; accordingly her compass is on a six foot high pedestal with a second placed high up the mizzen mast, this to lessen the effect of the iron on the compass's magnet.[16] I find this very interesting. I have not thought in the past of taking a compass aloft, but it may be a sound practice to do so in Australia, where I will not be familiar with landmarks, and where, in any event, there are such extraordinary distances to traverse.

I shall be very pleased to leave Suez.

Notes

1 Brown did not keep a diary of his journey to Australia. The style of the fictional diary produced in this chapter is drawn from the many examples found in Andrew Hassam's *Sailing to Australia: Shipboard Diaries by Nineteenth-century British Emigrants* (Melbourne: Melbourne University Press, 1995) 35 *et seq*.

2 *The Argus* 12 January 1858: 4. Coppin's account of his overseas trip given to the audience at the Theatre Royal, Melbourne.

3 Joseph Leach, *A Bright Particular Star* (New Haven: Yale University Press, 1997) chap. 14 *(unpaged)* Cardinal Book Electronic Edition ed.

4 James Barber, *The Overland Guide Book* (London: W.H. Allen, 1850) 41. Captain Barber sounded the warning regarding European travel in his guidebook, but since the publication was sponsored by The Peninsular and Oriental Steam Navigation Company, it may not be necessary to take the caution too seriously.

5 *The Argus* 12 January 1858.

6 William M. Thackeray, *Notes on a Journey from Cornhill to Grand Cairo*, Ch. XIV: 2. 13 April 2002. <http://www.bookrags.com/books/crhcr>.

7 Named in honour of the *danseuse?*

8 'Professor Anderson in Boston', *Gleason's Pictorial Drawing Room Companion*, 25 September 1852: 193. The command performance was given by Anderson at Buckingham Palace during 1843.

9 Thackeray 4.

10 Fly-boats were flat bottomed barges which could also be towed by horse or mules.

11 Thackeray 1.

12 Anna Sproule, *Port Out Starboard Home* (Poole: Blandford Press, 1978) 14.

13 Charles Waller, *Magical Nights at the Theatre* (Melbourne: Gerard Taylor Productions, 1980) 23. The Arnanoots or Albanian troops, brought to Egypt by the Ottoman rulers, were notorious for their blood-thirsty ill-discipline.

14 'Accident to the Emeu', *The Argus* 9 January 1858: 4.

15 Barber 5. The Egyptian Transit Company was responsible for the carriage of baggage, cargo and mail on the overland section while the European and Australian Royal Mail Company held the contract. When carriage of the Royal Mail passed to the P & O Steam Navigation Company, they organised their own overland transportation.

16 Charles James Griffith, *Diary written on board the Simla, 1858*, Ms 10875 La Trobe Aust. Mss. Collection, State Library of Victoria, Melbourne.

CHAPTER 14

The Simla

Thursday, 3 December. What a surprise! As the *Simla* got up steam for our departure yesterday, there were marched on board officers and men of the 69th Regiment. Clearly they had been in camp somewhere about Suez waiting for embarkation. I had no idea they were here. There are 300 soldiers in all, and they are bound for Galle. Now the ship's engines are rumbling in the depths, sending us through the Red Sea, but hopefully not too close to the shores of the Nubian Desert. There are two firemen who are natives of this part of the world. They are great-chested black Arabs, but splay-footed and bandy-legged. They tend to the boilers while we are under steam. For the 120 passengers on board there is a crew of 240. The main body is made up of Lascars, and there are a few Chinese. The Captain, his officers, including the Quartermaster, Engineer and Carpenter are all British as are the greater number of waiters. The Chinese catch the eye with their dress of sky blue nankeen trousers and jacket tight at the throat and loose everywhere else.[1] The *Simla* is narrower in the beam than the *Emeu* and I have been warned by fellow passengers who have made this trip before that she will, in consequence, roll a great deal in heavy weather. I suspect this may be another of those nautical tales told by the sterner travellers with a view of frightening the novice.

Saturday, 5 December. Lat. 12° 46′N – Long't. 45° 10′E. This seems a good time to commence a record of our latitude and longitude. I have obtained these figures from the first mate. We have this morning at 8am put into Aden. Or more correctly, docked near the fort of that name, which is three-quarters of a mile from the town. We will be here for two days taking on coal before we start for Ceylon – a trip that will take 10 days. Now that we are under weigh, Mr Coppin's disposition has taken on a more sanguine aspect, and he has invited me to play billiards with him this evening at a location known to him in the town. I am not an adept at the game, but I have tried my hand at it while I was at the Royal Gardens, and I thought it polite to accept the invitation.

There are many troops in Aden. It is a province of British India, so there are The Bombay Engineers, along with the 29th Native Infantry, and the Fort itself is guarded by the 57th Regiment Queen's Troops. Mr Coppin knows a number of the officers and we were invited to a club in the town. There are about five thousand souls living there. Apart from the British and Indian military and their families, the people are Arabs and Africans. Mr Coppin has a great sense of humour, and had us all in fits of laughter with witty tales, and many anecdotes of antipodean antics – I seem to have picked up on his alliteration. He is a man who can well hold his liquor, and he consumed, to my knowledge, at least six glasses of brandy. I had an ale. It is unusual for me to imbibe, but it seemed the proper thing to do in the circumstances. I did not disgrace myself at the billiard table, although I was a little unsure of how to properly score, Mr Coppin looked after me on that account and complimented me on having 'a good eye.' Mr Coppin had introduced me to the company as 'a veteran aeronaut' which was flattering, if inaccurate, and one of the officers had seen Mr Coxwell's experiment with his signalling from the War Balloon at London and spoke to me about it. It must have been an engagement arranged by Mr Coxwell himself, as I did not know the details of it. Mr Coppin has brought back on board two sheep and several monkeys.[2] I will sleep well tonight.

Sunday, 6 December. We are under weigh again. The coal that was taken on board has overflowed the bunkers and there is a large hillock of it sitting forward of the funnel and the main mast. Some of the second-class passengers have complained to the Captain, claiming the mound of fuel has restricted the space in which they can take their exercise. The Captain has countered by pointing out, since we burn a great quantity of coal when under steam, as we are at the moment, it will be but a short time before the deck is clear again. Mr Coppin remarked that he is more worried about the milch cows we have on board being upset for want of exercise, than he is concerning the mechanics' desire to promenade. I believe he intended his remark as a jest.

Thursday, 10 December. Lat. 12° 8 'N – Long't. 52° 0 'E. At daylight we were abreast of the island of Socotra. It is nothing more than a bare sandstone outcrop. There is no vegetation there whatsoever.

Sunday, 13 December. Lat. 9° 42 'N – Long't. 66° 22 'E. There was Divine Service today at 11am. The Captain read the Service from the Book of Common Prayer. An awning was rigged up over the quarterdeck, and this gave us all shade. About 100 attended and this congregation was composed mainly of passengers. I suppose the Lascars and the Chinese to attend to their own God at a time appropriate – whenever that may be. The Captain also led the singing of the hymns and the psalm. He has a good tenor voice.

The Nubian firemen are working apace to keep the engines stoked, and the coal is vanishing from the deck at an astounding rate, just as Captain Cooper claimed it would. I have remarked that these blackamoors sing wildly as they work, and now and then they will take to beating time on the ship's funnel. It gives off a musical sound but as it catches one unawares, there being no warning as to when the drumming is to commence, it can be alarming.

Tuesday, 15 December. Lat. 6° 1 'N – Long't. 80° 16 'E. We have put into Point de Galle, another fortified town, where the 69th. Regiment disembarked. They made a splendid display as they marched from the docks. Mr Coppin told me the regiment was know as the 'Ups and Down', and he sang a few verses from a rather ribald song – *The Maid of Tottenham* – I gather it had something to do with the regiment.[3] We will again remain in port for two days. At the moment we are accommodated in the town itself, in a large room divided into cubicles, with seven foot high dividers made of canvas and paper. The streets of this town are very narrow and crooked, but the houses are of handsome design, though the demeanour of the inhabitants does not match the grandeur of the architecture. The heat is very great, and on the advice of Mr Jones, who has been to Ceylon before, I purchased pyjamahs [sic] to sleep in. These are loose cotton trousers, which tie at the waist. They are in a loud pattern of red and white stripes, but are only for night wear, and will undoubtedly be cooler to wear than my night-shirt. The boy who sold them to me wanted 5/-, but I haggled with him, and in the end got them for 3/-.[4]

Monday, 21 December. Lat.1° 28′S – Long't. 85° 10′E. Today I witnessed a most curious sight. The crew were in a jovial mood. It seems, what with the actual time at sea and the delay in Suez, it is now time for the seamen to receive another pay. They have been thirty days without cash, having been paid in advance of the voyage, and that money now well spent. The sail-maker has concocted a 'horse' from scraps of canvas. It is not a very large beast, but it is weighed with old rope and bits of sandstone block. They paraded about the deck chanting, 'Poor old horse, you've got to die.' It was then hoisted up the yardarm before being dropped into the sea, with much hurrahing from the sailors. I am not sure of the actual significance of the ritual, but it provided a few minutes' welcome cheerfulness to an otherwise torpid passage of time.

Friday, 25 December. Lat.16° 4′S – Long't. 96° 51′E. What a curious experience it is to have Christmas Day at sea, and with the sun blazing down from a clear blue sky. Yesterday we passed Cape Leuwin. Mr Cole, the mate, tells me the word means lioness, and the location is named after Captain Abel Tasman's ship, *Die Leuwin*.

Thursday, 31 December. Lat. 38° 36′S – Long't. 119° 5′E. We are today at King George's Sound. It is a settlement of no more than 100 houses. There are soldiers and convicts here, and many natives. The aboriginals have their hair smeared with whale oil and their faces reddened with ochre and curious blackened cheeks and noses. They are clad only in small rug cloaks. They are a fearful sight, but they have set themselves to amuse us with their spears and boomerangs, and are content to be rewarded with a sixpence. Professor Anderson acquired a boomerang from one of the aboriginals and attempted to send it on a circular flight, but failed. I am sure Dr. MacSweeny would be able to pronounce on the aeronautical principle underlying the boomerang's capacity to return whence thrown. The crew is busily taking aboard a final load of coal from a hulk here. And this will power us on our final run to Port Phillip. Mr Coppin tells me we will be in Melbourne in five days time, but this has left all his plans in tatters, and he is very angry when he talks about it.

It is clear to me now that Eliza and the children will have already reached Port Phillip. Even Dean, who left in the *Shakespeare*, may well have made landfall by now. I only hope Eliza has been able to contact Mr William

Green. We were unsure of his address before we left, but have heard that he is a land agent in Carlton, so I sent word to him care of the post office, and told of my plans and our intended arrival. Hopefully, he will be known in the town by this time and it will not be too hard for Eliza to find him.

Thursday 7 January 1858. When we passed Queenscliffe, the Health Officer and Customs came aboard. They have telegraphed news of our arrival onward to Melbourne. I am sure that Eliza will receive this news and, thank God, we can all be united again. I realise how much I have missed her and the children, though there were many times the exciting prospects awaiting me and the very adventure of the ocean journey itself, put them at some remove from my daily thoughts.

Now I must leave off the writing of this diary and attend to the matters relating to our disembarkation, and the safe landing of the balloons. I would imagine that these stout colonials will be of a more reliable ilk than the Arabs who last handled the cargo and caused me such trepidation.

Notes

The readings of latitude and longitude are taken from Griffith's diary, and the dates from the *The Argus*, 'Shipping Intelligence', 9 January 1858.

1 Charles James Griffith, *Diary written on board the Simla, 1858*, Mss 10875, La Trobe Australian Mss. Collection, State Library of Victoria, Melbourne: 75.

2 *The Argus* 16 January 1858: 8.

3 The 69th Regiment of Foot were know as the 'Ups and Down' for obvious reasons. The traditional song, in which they are mentioned, dates back to the 1700s.

4 Griffith 91.

CHAPTER 15

Eliza's Journey

It was so strange that Charles and I should leave England on the same day, from different ports, on different ships. He did explain it all to me the reason why it should be so. There is no need for haste with respect to the children and myself, but Mr Coppin has given an undertaking to exhibit a balloon at his pleasure garden in Melbourne, to celebrate the New Year. I am sure I can manage on my own. Charles brought us to Plymouth on the Monday and then, having settled the passage for us, for which Mr Coppin had given the money, rushed off to get to Portsmouth for his ship. It was £15 for my steerage passage. Our ship is the *Swiftsure*, and it seems to be a fine vessel.[1]

Annie is such a help. She is, I think, enjoying being a little mother to Eugene, and Frederick has decided that he will have to be the man of the family, now that his father is not here. It was foggy when we sailed, so the last I saw of England was the harbour wall, shrouded in mist. Charles says that we will not be long away, just sufficient time for him to establish himself as a practical aeronaut. And while he speaks of exploring the interior of Australia, he has promised that we will return home within at least the next two years. I hope this will be the case, as I would not like our children to grow up in a strange country. Though the child I am presently carrying will be born in Australia, he will not be a native of that land, but an Englishman, like his father.

* * *

We are outside of the Bay of Biscay and I have begun to feel unwell. I thought at first it had something to do with my condition, but I then saw that it was a discomfort that was afflicting men also. One of the women in steerage, who has travelled by ship before, tells me it is seasickness. Of course, I should have known. I was warned that such a malady could be brought about by the rolling of the ship.

There was such a horrible creaking noise all the way through the Bay, that I believed something would break, or the bunks collapse. Frederick was in no way affected, and enjoyed himself thoroughly.

* * *

There was a great deal of rushing about on the deck this morning, with sailors shouting. Two young lads had been found in the hold. It seems they got on board when we were at Plymouth and had hidden themselves for the past week before being discovered. A young man, Mr Dodge, told me that he believed the last time the *Swiftsure* sailed to Melbourne ten such stowaways were discovered. I was told the young men will now have to work for their passage.

Steerage is quite crowded, but all the ladies help each other with the many chores. We are divided into messes, with twelve looking after each other, that is men, women and children. There are four young couples with me in our mess, and, having no children of their own, they fuss greatly with little Eugene, and I think they made Annie a little jealous.

There are many ways in which we keep ourselves amused: music, dancing, quoits, and for the men, boxing, while the children have great sport playing leapfrog. Last evening I declined the kind offer of one of the young men from our mess to partner me in the dancing; my condition is such that I thought it better not to be too active.

* * *

We are two weeks out from port and have just passed the Cape Verde Islands. It was exciting to see land, and some of the younger children on board thought we had reached Australia, and were very disappointed when told that this was not so. The island we passed was St. Antonio.[2]

* * *

It is the first Sunday of November and this morning we had the Captain, Mr Price, read the service. He conducted it from a table covered with the Union Jack, and was very impressive. Some of the sailors attended and sang the hymns very lustily.

Now that we have reached the tropics the weather is causing distress to some of the passengers. Many of the women have failed to bring light clothing, and one lady, Miss Logan, who is a Second Cabin passenger, has been having fits and cramps. I believe she is a lady who likes to be seen as fashionable and is in the habit of tight lacing. At noon-time today she went clean out of her mind, and ran about the deck and attempted to throw herself overboard. It took four sailors to restrain her. She then fainted and was taken below to the hospital.

* * *

Tonight was Guy Fawkes night. When it was dusk the sailors parading around the deck with a very strange Guy, which seemed to be dressed as the Captain, and there was a great deal of laughter and cheering from the crew and the men in steerage. About half a dozen Roman Candles were let off, which was nothing like the displays Charles organised at the Royal Gardens, or for Mr Coxwell, but it was sufficient to amuse the children, and then we all gave three cheers for the Captain, to show that there was no harm intended. (All the children are now safely tucked in their berths for the night.)

* * *

We crossed the line today. The children had been told Father Neptune would come on board and shave all the men, but Captain Price forbade it, which some said may have had to do with the Guy being made to look like him last night. If that was so it was a petty thing to do, depriving the children of a little enjoyment. But he seems to me a fair man, and I cannot imagine he would be so spiteful.

For a number of reasons it has become impossible to sleep below deck in the bunks. It is far too hot, and the children are distressed for want of fresh air, and on top of that we are terribly teased by fleas. Some of the men in steerage are in no way worried by either the heat or the pests, but many of the women have joined me on the deck with their children. I heard Mr Dodge suggest to the mate that if the men were allowed to smoke below deck then that would, like as not, oblige the fleas to flee. But I believe that was meant as a jest.

* * *

Now that we have moved beyond the Equator, the weather has started to cut up. We have endured a gale for the past three days. There are a number of theatricals on board who had been engaged by Mr Coppin to perform at his theatre in Melbourne, and they have entertained us on many an evening. Now in the rough seas, one of them, Mr Flint, a comic singer, has been thrown over and broken his arm. The ship has been sailing keeled over and everything is wet. The children are frightened. On two or three occasions the ship has been nearly jerked inside out and there are horrendous noise with knives, forks and plates rolling all about the place.

* * *

The storm has blown itself out. There is a general tidying of the bunks, and musical instruments have been brought out, and that has helped to make the work jolly. A number of ladies, who have been hired by Mr Coppin to present *Poses Plastique* at his theatre, came on deck to practice their performance.[3] I have never seen anything quite like it. Many of the sailors were lolling about in the riggings, and passing quite risqué comments. I sent the children below. This very night one of the intermediate cabin passengers was discovered in his cabin with his sister-in-law – he was found by his wife. She knocked up a great row and very quickly all the people on the ship came to know about it.

* * *

This morning, when the man who had been discovered last night with his sister-in-law came from his cabin, he was greeted by boos, hisses and groans. The men from steerage joined in, and called out such things as, 'Jam his tail', 'turn him out' and, 'Oh! The dirty old man.' His sister-in-law was greeted in the same way. The sailors are greatly amused by the whole thing, but with the Captain watching them, they have not joined in the sport.

* * *

Earlier today we came in sight of Western Australia. There was a general movement to get ready for landing, but the sailors told us not to excite ourselves and that it will be a week before we reached Melbourne.

* * *

It is six days since we sighted Western Australia, and this morning we passed into Hobson's Bay, where we are anchored. We have been told that tomorrow a steamer will come from Melbourne and tow us up the Bay.

* * *

We have brought our things on deck and, as I write this, we are waiting for a lighter to take us ashore. There is a very hot wind blowing, and a biting dust cutting into my eyes. Annie is very quiet, and Eugene is distressed, but Frederick is being very brave, and tells me not to worry and he will look after us. There is no sign of Charles.[4]

Notes

1 The *Swiftsure*, formerly named *The Fatherland*, was a clipper ship Carl Cutler, *Greyhounds of the Sea* (New York: Private Publication, 1930) 436.

2 Correctly the Island of Santo Antão.

3 *The Argus* 28 December 1857: 8. 'A highly talented Troupe of Artists for the presentation of Statuesque Groupings has arrived by the *Swiftsure*.'

4 Events in this chapter are broadly based on *A Journal of the Life of Frederick Joseph Dodge: Swiftsure Plymouth to Melbourne 17 October 1857–22 December 1857*. Mss 11530, Box 1768/1 La Trobe Australian Manuscript Collection, State Library of Victoria, Melbourne. Particular incidents in Eliza's 'diary' which are directly drawn from Frederick Joseph Dodge's journal are those which relate to the discovery of two stowaways; Miss Logan's attempt to throw herself overboard, and the passenger who consorted with his wife's sister.

CHAPTER 16

Introduction to Melbourne

Landing on the 21 December, and alone in Melbourne, Eliza quickly took control of her situation.[1] She came ashore with the children at Raleigh's Wharf and went with other passengers to Swanston Street, where they found lodgings. With the children she was billeted in a dormitory, and provided with a kind of litter on which to sleep. She was informed it was a 'stretcher'. She had never seen its likes before, but the children were excited by the prospect of sleeping on these strange devices, and their journey in steerage had inured them to cramped and crowded quarters.[2] Inquiries after William Green had established that he was away from town; that is if it was Mister Green, the land agent, she was seeking. Eliza was not sure if it was.

By the time Brown arrived on the *Simla*, Eliza had established the family in a modest rented house at York Place, Emerald Hill, on the south bank of the Yarra, across the Melbourne and Hobson's Bay Railway line, from the docks. Her task had been made a little easier by the precipitous departure of the previous tenant. He had taken fright in consequence of a number of house fires in the adjacent street, and had moved on for what he considered the relative safety of the goldfields.[3]

After their arrival, and satisfied that the precious balloons were safely ashore, Coppin gave Brown a cursory introduction to Melbourne. 'It is a good town, quite English, and you will scarcely fancy yourself away from home.'[4] Brown had thoughts to the contrary. The light was glaring and harsh on his eyes, the hot wind and dust was close to intolerable. He wondered how it could be considered in any way practicable even to attempt to get a balloon aloft in such conditions, but Coppin did not give him time to discuss these misgivings. 'Take yourself to the Gardens on one of the steamers, the *Gondola* or the *Pride of the Yarra*. Tell the helmsman who you are, and he'll look after you.'[5] It was an unsatisfactory introduction to his new employment.

Dean had arrived on the *Shakespeare* before anyone else and promptly vanished.[6] But no doubt, with word of Coppin's arrival now spread about the town, he would make an appearance. Coppin was very quick to make public the news of his return. He had made a promise to his patrons, and he would now have to make his excuses. He placed an advertisement in *The Argus*, announcing his arrival from England, and proposing to explain at the Theatre Royal, on Monday 11 January, just what he had achieved in regard to the engagement of novel entertainment for Melbourne.

On the night, the theatre was packed for Coppin's return appearance. He had taken Brown along and stood him in the wings, just in case he chose to introduce the 'veteran aeronaut'. Before he got to the business of announcing the talent he had – and had not – engaged for the coming season, he entertained his audience with a comic dialogue, presented in the character of Jem Baggs from Mayhew's *The Wandering Minstrel*. Then, still in the character of Baggs, he launched into a parody of Stephen Forster's *Old Dog Tray*:

Old dog Tray! always hungry
Meat cannot drive him away:
With his tail 'druv' in behind,
Neither gentle nor yet kind
Is that hungry dog, that old dog Tray.[7]

Brown was surprised to recognise the song, but quickly placed where he had heard it – those months before at the Winchester. When Coppin was vociferously encored, he obliged with *Villikins and his Dinah*. Amid cheering and hoots of laughter he concluded, with a large handkerchief clapped to his eyes:

And all of you parents both miserable and grey
Beware of my warning, take heed what I say.
Don't never try to wed your daughter off
To one she never clapped her eyes on,
Think of Villikins and Dinah, not forgetting the cold poison.

This rapport with the audience established, it was time for him to give the bad news. Coppin reported on his failure to secure the services of Charles

Mathews, Madame Celeste, Miss Cushman and Mr and Mrs Williams. This announcement was greeted with good-natured groans. Then, with a flourish, he declared he had procured the services of Professor Anderson, The Wizard of the North. The orchestra greeted this news with a sustained chord, and the audience cheered lustily. Coppin was moved to declare theatrically that the reception he had received convinced him to leave off his wanderings and to 'strike up' for the remainder of his life in his adopted country.[8] This sally was greeted with more gratifying applause. As he had determined, Coppin now laid the blame for all his troubles at the foot of the European and Australian Mail Company. They were solely responsible for his failures. They had been tardy in delivering him to England, and it was their bad management, he claimed, that had left him so many weeks late in effecting his return. Curiously, he made no comment on the bad navigation on the part of the *Emeu*'s skipper. But, despite all such tribulations and hindrances, he had brought out 'fifty minor nobilities', and Professor Anderson was following. He had also employed two aeronauts, and brought out two balloons. Granted, he had originally intended that one of these craft make a pioneering flight on Boxing Day, but better late than never, and the date for Victorians to witness the colony's first aerial ascent was set for 1 February, at the Cremorne Gardens.

Coppin's announcement of the event appeared in the newspaper the following day. Brown quickly went to the Gardens to conduct a survey of the grounds. Cremorne Gardens was two miles out from Melbourne. It comprised ten acres of land, laid out as a Botanical Garden. There was a collection of Australian animals and birds, a hotel, drinking booths and a dance floor. The establishment was immodestly billed as 'The Vauxhall of Victoria'.[9] The City of Canton was represented in a giant outdoor panorama, set on one side of a large artificial lake. There was a theatre for concerts, vaudeville and ballet. Coppin's partner, Gustavus Brooke, the actor, made occasional appearances there.[10]

One of the problems, which became immediately apparent, was that the gasworks which Coppin had at Cremorne simply did not have the capacity to produce gas in sufficient quantities to provide for the needs of the Gardens and also allow for the inflation of the *Australasian*. It required 30,000 cubic feet of gas to ensure reasonable buoyancy for the balloon.

Brown calculated that he would be hard pressed to inflate even the smaller *Tavistock* using the facility at the Garden. He would need to call on the services of the City of Melbourne Gas and Coke Company. Unfortunately, the gasworks were situated across to the west of the town at Batman's Swamp and a means of acquiring an extra supply of gas would have to be organised with some urgency, if the flight of the *Australasian* were to take place by the first of February. Although the newspapers carried reports about the balloon, no notice had yet appeared advertising the ascent. Brown had seen a report of the arrival of Coppin and the proposed appearance of the Great Wizard Anderson, but nothing regarding himself as a newly arrived pioneer aeronaut has so far appeared in print.

However, Brown's feelings towards Anderson had mellowed, and he did not grudge the man the publicity he'd received. Towards the end of the voyage Coppin had told him how the Great Wizard had been left penniless following the collapse of the Royal British Bank where his life savings had been lodged. He had equipped himself for the Australian trip using borrowed money.[11]

The preparation for the balloon ascent, and the proper promotion of that event now required urgent attention. Brown, apprised Coppin of the problem regarding the supply of gas and was instructed to approach the Gas Company's engineer, Alexander Kennedy Smith, and seek his help. Smith agreed to provide some of the gas required at a cost of 22/6d per thousand cubic feet. The balloon would need to be topped-up at the Gardens. Matters generally had fallen behind time at the Gardens. Overseeing of the opening of the Summer Season had been left to Gustavus Brooke, while Coppin was away in England.[12] But he had other things on his mind, such as the arrangements to be made for the laying of the foundation stone for a new theatre in Ballarat.

Fortunately, there was a ferry across the Yarra, reached by a causeway across the swampy land from the foot of Emerald Hill, and that gave Brown ready enough access to the gasworks, there to make the necessary arrangements. A quick inflation was a prime consideration, as even at the best of times the balloon inevitably lost gas through leakage. The slower the filling, the greater the loss. It was desirable that the pipes used to feed the gas into the craft should have the widest possible diameter, and Brown had already noted that the pipes at Coppin's private gasworks were very narrow.

The first public notice of the intended aerial ascent finally appeared in *The Argus* on Wednesday 27 January 1857. It was less than satisfactory, stating simply that 'first particulars' would appear the following day. The second promotion was a far more effective presentation:

> The first BALLOON ASCENT in the Australian Colonies will be made by Messers. Brown and Dean in Coppin's Australasian Balloon on Monday 1st February at 5 o'clock. Mr Brown is a pupil of the celebrated English aeronaut Mr Henry Coxwell, and Mr Dean is well known in England having made 119 ascents from Vauxhall and Cremorne gardens in London, and various towns in England. The Inflation at 2 o'clock. An Extra Grand gala in the evening. Admission 5/-. Children half-price. A number of pilot balloons will be discharged during the day. The gas will be supplied by the MELBOURNE GAS COMPANY under the direction of Mr A.K. Smith.[13]

The *Australasian* was unpacked and checked. The valves were adjusted and the netting examined. The car was attached and the wooden toggles – a Brown invention – put under pressure to make sure that all was sound and would take the weight of the two aeronauts along with all the necessary ballast, ropes and grappling irons. Everything was found to be in order. Dean, who had emerged from somewhere in the city centre where over the previous days he had been sampling the town's alcoholic delights, was sufficiently recovered from his expedition to assist in the preparations. It seemed as if Coppin, Brown and Dean were about to make Australian aeronautical history. And Eliza was about to give birth to Brown's third son.

Notes

1 *The Argus* 23 December 1857, sec. Shipping Intelligence: 4.

2 *A Journal of the Life of Frederick Dodge: Swiftsure, Plymouth to Melbourne 17 October 1857 – 22 December 1857*. Mss 11530, Box 1768/1. La Trobe Australian Manuscript Collection, State Library of Victoria, Melbourne: 27.

3 'Fire at Emerald Hill', *Argus* 19 December 1857: 4.
4 Andrew Hassam, *Sailing to Australia* (Melbourne: Melbourne University Press, 1995) 204. An extract from the diary of J.W. Reeves, 13 February 1857, commenced on board the *Queen of the West*.
5 Agnes Bell, *Melbourne: John Bateman's Village* (Melbourne: Cassell, 1965) 83.
6 The *Shakespeare* arrived at Melbourne on 19 January 1858. *The Argus* 21 January 1858, sec. Shipping Intelligence: p. 4.
7 *The Argus* 12 January 1858: p.4. The reporter mistakes the song as being a parody of *Campbell's My poor dog Tray/The Harper*. Frederick Robson was well know for his interpretation of Jem Braggs, and had virtually made *Poor dog Tray* and *Villikens and his Dinah* his theme songs in England. Coppin had staked his claim to the songs in Australia.
8 'Theatre Royal', *The Argus*, 12 January 1858: p. 4.
9 'Amusements', *The Argus* 24 March 1856: p. 8.
10 W.J. Lawrence, *The Life of Gustavus Vaughan Brooke* (Belfast: W & G Baird, 1892) 189.
11 Milbourne Christopher, 'Anderson in America', *MUM* Vol 44. December 1954.
12 Lawrence 188. 'I am very busily engaged with [Cremorne Gardens] as we open on Monday next, November 23rd [1857].' Brooke.
13 *The Argus* 28 January 1858: 8.

CHAPTER 17

A Balloon on the Ballaarat

Brown had little time to think of anything but getting ready for the ascent at the Gardens. Dean had appeared after an absence of three days and, though looking very much the worse for wear, he was sober. Eliza Brown gave birth to her fourth child, a son, Herbert. Ten-year-old Annie and the two boys helped their mother cope with their new brother as best they could. Brown was so embroiled in readying the *Australasian* that he neglected to have the birth registered. Unbeknownst to Coppin and Brown there were similar preparations in hand in another part of the colony.

Sir Henry Barkly was twelve months into his governorship of the Colony of Victoria. He had been appointed in November of 1856, and had landed in Melbourne with his wife and family on Christmas Eve of that year.[1] His arrival was earlier than expected and at an inconvenient time for the quick arrangement of any grand welcome. Sir Henry was a man whose manner was less formal, and his style more relaxed, than that of his predecessor, the autocratic Sir Charles Hotham, and at that time he viewed the absence of any fanfare as a blessing, since his wife was not in good health.[2]

Now, twelve months into his governorship – and some nine months after the tragic death of his wife – the Governor announced his intention of going to Ballaarat and laying the foundation stone of a new Wesleyan chapel.[3] *The Argus* reported that:

> the various public bodies have exhibited much zeal, and have entered upon the business of making arrangements for giving His Excellency a becoming reception, with an eagerness and determination which says much for their loyalty and for our governor's popularity.[4]

The Municipal Councils of Ballarat East and Ballaarat [*sic*] West both enthusiastically forwarded arrangements. Ballarat East declared 19 January

1858, the first day of the Governor's visit, a public holiday.[5] Ballaarat West followed suit. Triumphal arches were erected, one at the Yarrowee Bridge and another at the junction of Sturt and Lydiard Streets. A band of twelve musicians was engaged to play in front of the council chambers during the Governor's visit.[6] The inhabitants of both municipalities were called on to decorate and illuminate their houses, a task they undertook with alacrity.

That the Ballaarat was working itself into a lather of enthusiasm over the Governor's impending visit was not something that had escaped the notice of William Green. 'Bill', as he was happy enough to have himself called, had achieved over one hundred balloon ascents in England before he departed for Australia. Son of William Green senior, an aeronaut, he had frequently gone aloft in England as a crew-member to his father and his uncle, Charles Green, the veteran flyer.[7] He had also flown solo on many occasions. Though he had opted to make his living as a land agent in his adopted country, he had not given up entirely on aeronautical endeavour.

Melbourne and Sydney had seen a number of attempts by itinerant aeronauts to take to the sky. None had been successful, and a few had been disastrous. In December of 1856 Pierre Maigre had so mismanaged his attempted ascent in a Montgolfier balloon from the Domain in Sydney that a fatality had resulted, and a riot ensued.[8] Green had sought to assist another French aeronaut in Melbourne, but nothing had come of the attempt. The Melbourne newspapers had kept him abreast of Coppin's latest news. He knew that his old acquaintance, Charles Henry Brown, was bound for Melbourne with a balloon, and was well aware of Coppin's avowed intention to have Australia's first aerial ascent from the Cremorne Gardens; it was a matter of general knowledge – Coppin and Brooke in conjunction having ensured it would be so. Coppin's tardy return from Europe, however, had fired Green with the notion of pre-empting the entrepreneur's aerial aspirations. In fact Green and Coppin knew each other and were on bad terms as a result of Green's earlier attempts to aid the French aeronaut. Such an undertaking had been seen by Coppin as an intrusion into a sphere of public entertainment that he regarded as being his singular domain, and he did not take kindly to trespassers. This lingering ill-feeling had acted as a spur to Green in his formulation of a plan to anticipate Coppin's Cremorne adventure.

Now the Governor's proposed visit to Ballaarat had focused attention on that part of the world. Green, over some time, had manufactured a gas balloon. He always had an eye to his own eventual return to the air, and with his craft stowed in its basket, he had it slung on the baggage rack of a Telegraph Line mail coach. Having paid his thirty shillings, he set off on his over-night trip to Ballaarat.[9]

Gustavus Brooke, Coppin's partner had also been intent on taking advantage of the Governor's visit to the twin municipalities. Plans for the construction of a new theatre to be named The Theatre Royal and located on Sturt Street in Ballaarat West, had been in train for some time. And, with the attention of Melbourne now focussed on the Ballaarat, Brooke hit on the idea of a foundation-stone laying ceremony as an additional attention-catcher. The Governor would arrive on Tuesday, 19 January, so Brooke decided that the Wednesday would be ideal for his ceremony, and he made arrangements accordingly. He would ensure a good coverage in the press. He was adept at handling newspapermen, and held the view that when charm and cajoling failed with the journalistic fraternity liquor was an apt substitute. Brooke himself was an enthusiastic bibber. *The Star* quickly announced the commencement of the theatre project, and noted that the stone-laying ceremony was to be 'surrounded with as much *prestige* as can conveniently be given to it'.[10] And Brooke had no intention of staying out of the public gaze until the ceremony, however. He undertook a twelve night engagement at the Charlie Napier Theatre, where, under gas-light, he performed in *The Hunchback* and *Othello*.[11]

In the planning of his foray, Green had to ensure that a suitable supply of gas was available. In Melbourne the Gas and Coke Company were the main providers, but Ballaarat had no such large supplier. Henry Courtis from Castlemaine had, however, developed a system of gas manufacturing that used oil and gum leaves, and such was the success of his process that in October 1857 he had opened a plant on the low ground east of Lydiard Street. From there he provided gas to Bath's Hotel, the Golden Fleece, the George Hotel and others.[12] Those places that were not directly supplied by Courtis contrived to copy his production method. The Charlie Napier Hotel was one such establishment that undertook the production of gas from animal fat. The system was quite successful. There were two hotels adjacent to the gas-works that were supplied by Courtis. These were Underwood's *Unicorn Hotel* and Laidler's *Robin Hood Hotel*. Green made an arrangement with the hotels to act as his sponsors, and Courtis could generate sufficient additional gas to inflate Green's balloon. The balloon would go up on two consecutive days, the first being on the nineteenth of the month, when the Governor arrived. Green would, of course, carry passengers aloft.

* * *

The performance of *The Hunchback* at the Charlie Napier Theatre on the previous night, Monday 18 January, had been a huge success, and the post-performance party was a boisterous affair.

'Jasus,' breathed Gustavus Brooke into his pillow. He turned slowly onto his back and groaned. 'Jasus.' A tangle of bed clothes was caught about his neck, and for a moment he thought he was being attacked. He struggled and thrashed at the aggressive bedding, and then lay back, gasping for breath. He turned his head to one side, and there on the washstand spotted his life-saver. The bottle of Jameson sat there. It wasn't going to come to him; he, perforce, must go to it. He was unsure if his legs would cooperate in effecting even the shortest journey at this hour of the morning. A thought struck him. Had he still got legs? He groped under the sheets. He realised that, the blinding pain in his head notwithstanding, he was still not sober.

He moved his feet, and slowly inched his heels towards the side of the bed. He could just make them out as they emerged from under the covers towards the end of the bed. So, he still had legs. Slowly he eased himself upward, into a semi-crouched posture. 'Concentrate,' he told himself, as he teetered on the side of the bed. Not quite knowing how he had managed the manoeuvre, he found himself clutching the marble washstand. The first swig of the whiskey acted as no more than a mouthwash. The second swallow did the trick. The pain did not abate, but he could now see clearly, and not as before through a mist. Cautiously he made his way to the draped window, his feet catching in his clothing that was strewn about the floor. He pulled back the curtains, and recoiled.

'Tearing blue blazes! What is God's name is that?' Bobbing above the rooftops and reflecting flashes of the bright morning sun as it twisted, was a giant yellow orb. Brooke stumbled to the hotel room door and yanked it open. 'Boots! Boots!' he bellowed down the stairwell, and overcome by the exertion, staggered back into his room and collapsed onto the dishevelled bed. After a minute or so there was a faint tap on the half-opened door.

'It's me, Sir, boots, Sir.' called a voice from the corridor.

'Come in, boy,' called Brooke imperiously. He had pulled his Garibaldi jacket about his waist, covering the drawers in which he had slept. The 'boy' who came in as commanded was a weedy man of indeterminate age. Brooke waved a hand from the bed.

'Bring me that,' he demanded, indicating the bottle of whiskey, and added, as an afterthought. 'And find me a glass, like a good chap.'

The boots plucked the bottle from the washstand and fossicked in the cupboard, finding a glass stowed behind the unused chamber pot. Seeing that convenience caused Brooke to wonder for a moment just how he had survived the night, unrelieved. But he quickly put the thought from his mind.

'I've brought you a paper, Sir,' said the emaciated boots, handing over the bottle and glass and fumbling under his waistcoat to produce a copy of *The Ballarat Times*.

Brooke tucked the bottle and newspaper under an arm and propped up the pillows and bolster behind him. When settled, he uncorked the Jameson and poured himself half a glass. He raised it in a mock salute towards the boots, who was at the window tying back the curtains.

'And just what is that monstrosity out there?' he inquired of the servant, as he settled the newspaper on his lap and took a deep draught of the whiskey.

'Oh, that's in the newspaper, Sir,' remarked boots unhelpfully. Brooke cocked a quizzical eyebrow at him. 'In the middle page,' explained boots. 'It's a balloon.'

'Holy Jasus!' cried Brooke, scrabbling off the bed, though taking care not to spill his whiskey. 'Let me see again.' His vision now sharpened to a degree by the fresh surge of alcohol into his system, he saw clearly that the heaving yellow sphere was indeed a balloon. 'Get my clothes, man,' he directed, 'I've got to put a stop to this.' The boots moved towards the wardrobe to comply with Brooke's direction. 'On the floor, you fool,' shouted Brooke. 'My clothes are on the floor.' As the boots fossicked about the floor, Brooke spread out the newspaper on the bed. There indeed was the notice: 'A balloon will be let off from Mr Laidler's Robin Hood Hotel, Plank road on Tuesday evening at nine o'clock. Any party for an aerial flight must be in time.'

It was no great distance from Bath's Hotel to *The Ballarat Times* office, and Brooke strode it out in a purposeful manner, surprisingly feeling a little better with each step. The anger he felt had an invigorating effect on him. He threw open the door of *The Ballarat Times*, startling the few patrons who were crouched over the reading desk.

'Where is Moloney? Where is that gobshite editor?' he cried. He was aware of the dramatic nature of his entrance, and was quietly pleased with the effect. 'Get out here, you scoundrel,' he called towards the back of the office. In response a voice called out, 'What is all this hub-bub?' Joseph Dunne, one of the proprietors of the newspaper, emerged from his office at the rear of the building.

Dunne was a well-known barrister as well as being a newspaperman. He knew Brooke well, both as an actor and as a fellow Irishman. He had attended at the Charlie Napier the previous evening. 'Why, Gustavus, you are out and about bright and early,' he remarked jovially.

'The very man,' exclaimed Brooke. 'Get yourself back into that office! You and I have some talking to do.' He waved Dunne back with his rolled copy of *The Times*. The alcohol was now beginning to work favourably on his system, and he felt at his imperious best. 'This will not do,' he shouted, whacking down the paper on Dunne's desk.

'What won't do?' asked Dunne calmly, retrieving the paper and shaking it open. He could smell the whiskey on Brooke. He was also well aware of his weakness for liquor, and was prepared to make allowances for his erratic and flamboyant conduct.

'There!' said Brooke, jabbing at the publication with his finger. 'Have a look at that advertisement on the middle page.' Dunne spread the paper to the appropriate page. 'See that,' continued Brooke. 'A balloon ascent, today. We can't have that. My partner is going to have the first ascent in the colonies in a few weeks. We can't have anyone getting in ahead of us on that little lark. It will have to be stopped. That's a job for us.'

Dunne lowered the paper and looked at Brooke. 'Us?' he asked. Brooke ignored the jibe, and Dunne shrugged. 'I would suggest that you are better positioned to deal with the situation than I,' stated Dunne. 'You are friendly with Laidler, are you not?'

Brooke paused before answering. 'Yes,' he conceded, 'I know him well enough.'

'So, there you are. Go and have a word with him about this balloon, and I'm sure the whole affair can be quickly, and quietly, deflated.' Dunne could not resist a smile at his own wit. Brooke missed the point. 'But before you go, Gustavus, I should warn you that tomorrow's edition of the papers carries an advertisement for another ascent.' He picked a proof from his desk. 'Here, listen to this: "Green's Balloon Extraordinary will be started from the Unicorn Hotel this morning at 11 o'clock"', he read. 'And this page and others have already been run. The only material remaining to be set will have to do with the Governor's functions today.'

'What are we going to do?' enquired Brooke despairingly.

'My dear Gustavus, I am going to see to the production of my newspaper,' stated Dunne firmly. 'But I would suggest that should you wish to do anything towards impeding Mr Green's proposed balloon ascent, then follow my suggestion and take the matter up with Messers Laidler and Underwood.'[13] Brooke had slumped in a chair and was holding his head in his hands. Dunne felt a twinge of sympathy for his countryman.

'Look, here's what I'll do,' he continued, in a conciliatory tone. 'Should your opponent, Green, actually get his balloon off the ground, I will not give the occurrence any copy. Whatever you and Mr Coppin think of such things, I have no time for foolishness of that nature, and Lord only knows the Governor will keep me busy and provide me with as much material as I need to fill two editions. As to what the *Star* gets up to, well, that you will have to attend to yourself.'

Brooke slowly rose from the chair, and extended his hand. 'Thank you, my dear Dunne. That is most kind of you, and I will not forget it.'

Brooke needed another drink, but at least he could see that the matter of Green and his balloon was taking on the aspect of being a time-consuming bother rather than an insoluble problem. The two hoteliers would be loath to lose his goodwill, and *The Ballarat Star* would certainly not entertain the prospect of losing the future advertising from the Theatre Royal. Yes, all in all, thought Brooke, the matter could be dealt with – it was time for another drink.[14]

Notes

The name Ballaarat is derived from two aboriginal words, *balla*, meaning to rest or recline on an elbow, and *arat*, a place. During the mid-nineteenth century Ballarat had two municipal councils. Ballarat East preferred to use the 'short' version of the name, while Ballaarat West opted for the 'extended' form. Diggers would make reference to 'being on the Ballaarat.' When a new unitary city council was formed in 1894 the single 'a' version of the name was adopted. Curiously the newspapers serving the dual municipalities used the spelling 'Ballarat' in their banners.

1 'Sir Henry Barkly', *Australian Dictionary of Biography*, Vol. 3 (Melbourne: Melbourne University Press, 1969) 95.
2 William Kelly, *Life in Victoria*, vol. II (Kilmore: Lowden Publishing, 1977) 350.
3 *The Ballarat Star* 12 January 1958: 3.
4 'The Governor's Visit to Ballaarat', *The Argus* 19 January 1858: 5.
5 'Minutes of Meeting 12 January 1858', *Minute Book – Municipal Council Ballarat.*
6 'Minutes of Meeting 15 January 1858', *Minute Book – Municipal Council Ballarat.*
7 Coxwell to Brown, 1 November 1858, *Correspondence*: 295.
8 'The Balloon Monster Swindle', *Bell's Life in Sydney & Sporting Review* 20 December 1856: 2.
9 An overnight coach service to the Ballaarat had been running since early 1856. It was also possible to travel on a day journey by steamer and coach via Geelong. *Argus*, 2 January 1857 carried an announcement of such a service.
10 *The Ballarat Star* 12 January 1858, sec. Local & General News: 3.
11 Theatrical tradition has it that Brooke's inability to take to the stage in Sunderland in 'Othello' – having been discovered blind drunk amongst the scenery –

provided his understudy, Henry Irving, with his opportunity to play his first professional leading role.

12 'Gas', *The Ballarat Star* 14 October 1857.

13 Underwood was the proprietor of the Unicorn Hotel. *The Miner & Weekly Star* 29 January 1858: 75.

14 For a fuller treatment of this episode, see Terence O'Neill-FitzSimons, 'A Balloon on the Ballarat', *Victorian Historical Journal* Vol. 71. No. 1 (March 2000): 27.

CHAPTER 18

Aloft

When *The Ballarat Times* was circulated on Wednesday, 20 January 1856, it carried a notice to the effect that Gustavus Brooke would lay the foundation-stone of the Theatre Royal at the building site on Sturt Street. The ceremony was set for three o'clock and was calculated not to clash with any of the Governor's engagements. The attendance of all was solicited, especially 'shareholders, Ladies and Gentlemen of the Dramatic Profession, Members of the Garrick Club, and Literary and Dramatic Society'. A special platform was erected for the exclusive use of the ladies.

As Dunne had warned, the advertisements for Green's balloon ascent also appeared in *The Ballarat Times*. And, like as not to Brooke's annoyance, it was placed directly under the Theatre Royal announcement.[1] At least he had secured the co-operation of the two hotel proprietors and successfully grounded the proposed balloon ascents.[2] So, despite his present irritation, the overall result had been a satisfactory one, and the whole incident could be made into an amusing story to relate to Coppin, when he got back to Melbourne.

At the appointed time hundreds of spectators gathered on Sturt Street to watch their favourite thespian in action, and Brooke made a good production of the affair. When the large foundation stone was in position, he jumped onto it and launched into a vigorous speech. He pronounced the people of Ballarat to be 'men of marvellous energy'. This went down very well with the crowd, and the band in attendance broke into *For He's a Jolly Good Fellow*.[3] The formal part of the proceedings having been disposed of, the official party and as many of the spectators as could be accommodated retired to the Clare Hotel. There numerous toasts were drunk. William Brown, the Chairman of the Theatre Committee, proposed a toast to the Press. This was enthusiastically replied to, and at some length, by Bailey of *The Ballarat Star*. The theatre project was well and truly launched.[4] On the heels of all the festivity in Ballarat that day, Green's balloon vanished without trace, or mention, and Brooke shortly took himself back to Melbourne in right good humour.

Unaware just how close his aerial project had come to being trumped in Ballarat, Coppin continued to make preparations in Melbourne for the *Australasian* to be taken aloft from the Cremorne Gardens on Monday, 1 February. There was a belated flurry of advertising activity. Brown had suggested to Coppin that one of Hampton's old promotional ploys be used to promote the event – the offering of a reward for the return of the balloon to the Gardens from wherever it should descend. Coppin promptly posted notices advertising a £5 bounty. He was also urged to promote the 'scientific' nature of the ascent, and the role of aerostation in the future exploration of the interior. This he did, paradoxically posting advertisements in the *Amusements* column of the papers. One advertisement posed the arcane question, 'what effect will the purity of air have upon Coppin's Balloon?'[5] Another advertisement appearing on the day of the ascent put the more pragmatic proposition:

> It is believed that the interior of this colony [sic] will be explored by means of a Balloon. The event of this day is therefore rendered doubly interesting as it will illustrate the practicality of UNLOCKING THE LANDS yet unknown to the satisfaction of all classes. Cremorne Gardens at 5 o'clock.

The Argus picked up on this line and wrote that:

> the first ascent of a balloon in these colonies is an event of considerable interest, not only for the general public but for men of science, as the experiment which is to be made this afternoon by a couple of experienced aeronauts may assist in determining to what extent aerial navigation can be made subservient to the exploration of this continent.[6]

Finally, after frantic preparation, all was ready. *The Age* on the day of the announced ascent suggested that the event 'cannot fail to draw immense numbers to this favourite place of entertainment'.[7] The balloon, partly inflated at the Gas and Coke Company yards, was loaded on a cart.

Attended by a number of workmen it made its way slowly through the crowds towards Cremorne. The ten acre Gardens were ringed on three sides by a stout bluestone wall, the Yarra River to the South acting as a defence against the incursion of non-paying guests, and a single landing-stage providing access for those using any of the steam-boats from Princes Bridge. Because of the press of people there was some delay in getting the balloon into the grounds. It was becoming evident that many of the spectators were not prepared to pay the increased entrance fee that Coppin had introduced for the event, and were taking up whatever vantage point they could find outside the Gardens.

Brown supervised the unloading of the balloon and quickly connecting the machine to Coppin's gas-making plant. The final inflation began at 2 o'clock. As Brown had predicted, the narrow gas pipes through which the gas was being fed resulted in a very slow inflation. It was clear that the machine would not be ready to ascend on time. Notwithstanding all the precautions he had taken there was now a serious doubt in Brown's mind that the balloon could be inflated to its full 31,000 cubic foot capacity. Forty minutes after the time scheduled for the ascent the *Australasian* still bobbed flaccidly at its moorings, with insufficient buoyancy to lift the two aeronauts aloft.[8] Joseph Dean early on had ignored the clear need for as many hands as possible to assist with the task of steadying the great swaying sphere of the balloon, and had stood nonchalantly to one side; now he ensconced himself in the basket and made minor adjustments to the rigging. Still not entirely satisfied with the craft's buoyancy, Brown had the gas-pipes disconnected from the craft and climbed into the basket.

* * *

'Could you not have made yourself a little more useful?' snapped Brown. Dean ignored the question and, taking some bags of ballast from the floor, hung them over the side of the basket. 'Dean, we simply do not have enough lifting power!'

'Yes,' drawled Dean. He glanced up at the towering sphere. 'We're losing gas, you know. Did you check the valve?' A band, not more than twenty yards away from the balloon, was giving a lusty rendition of the popular dance, *Pop Goes the Weasel.*

'Of course I checked the valve,' barked Brown. 'But must I do everything myself? It's the heat. It's too hot and we have not taken on sufficient gas. I told Coppin the pipes were too small.'

'Perhaps you'd like to explain that to the crowd here,' said Dean, with a sweeping gesture. 'Though I get the impression there are more without than within.'

'You've been drinking,' said Brown. Loquaciousness on Dean's part was a sure sign that he had been imbibing.

'Not a drop,' retorted Dean. 'Just intoxicated by the prospect of being the first aloft in this colony.'

'Have some sense, man. Unless we do something quickly, this balloon is not going to leave the ground with us aboard.'

'I agree,' replied Dean with a smugness that enraged Brown. 'This craft will not lift two of us. So I invite you to step down and I will ascend.'

'No you will not. Not in my balloon you won't!' Brown was beside himself with rage. Here was this wire-maker trying to deprive him of the honour of taking his own balloon aloft.

'My dear Brown, if I were you I wouldn't let Coppin hear you make such a claim. Your balloon, indeed.' The band was now playing the old number, *Aerial Polka*, with the bandmaster anxious that they had almost exhausted their repertoire.[9] 'Anyway, this is likely to be a tricky flight in the circumstances, so I, having more experience than you, will attend to it.'

This unctuousness was too much for Brown. He jumped from the basket and, pushing past the men now holding the tether ropes in anticipation of an ascent, went urgently in search of Coppin. He did not prove hard to find as he was energetically making his way towards the balloon.

'What is going on, Brown?' He took the aeronaut's arm and turned him back towards the balloon. 'Why are matters so delayed?'

'As I warned, Mr Coppin, it is the gas. The narrow pipes here cannot pump it through quickly enough,' said Brown by way of explanation. 'We are now close to losing more gas than we are taking in. It may be a valve, but I suspect it is in consequence of the great heat. The varnish has dried and the skin is porous, and the pipes ...'

'So, what is all that?' interrupted Coppin. 'Are you telling me the balloon will not fly?'

'Not with two of us aboard,' said Brown. 'And Dean will not give way to me.'

'This is not the time to …' started Coppin, but was cut short by a boisterous roar from the crowd. Dean had anticipated any decision by his employer and had ordered the release of the tether ropes. 'Well, that seems to have settled that,' said Coppin, and walked away. Dumbfounded, Brown watched his balloon climb into the Australian sky.

* * *

The *Australasian* had lumbered into the air, and with Dean already ejecting ballast the craft just cleared some trees on the North East perimeter where caught by an easterly air-stream it began to gain altitude. At a height of three thousand feet the *Australasian* drifted towards Brunswick. After twenty minutes Dean decided that he had done sufficient in the way of aerial adventuring. He was well away from Cremorne, the spectators had seen what they had come to see, and it was now time to return to earth. But there was little Dean could do to control his craft. He had jettisoned most of the ballast in getting the *Australasian* airborne and while he could use the gas valve to bring the machine down, he had no means of slowing its descent. The machine came down heavily on the Sydney Road, and – as Dean noted cheerfully, before being shaken violently in the uncontrolled landing – very close to the Cambridge Arms. Bruised but elated, he stood in risk of more serious injury by being caught up in the rowdiness that erupted as various carters, wagoners and hackney drivers pressed their claim to return the aeronaut to Richmond and claim the £5 reward offered by Coppin.[10]

For all the problems with the inflation, the *Australasian* had come through her maiden flight unscathed. But the day had been a disaster for Brown. He felt angry and cheated at having lost the opportunity to pass into history as the man who had made Australia's first aerial ascent. Nor was Coppin entirely at ease with the results of the day. The event had failed to provide him with the anticipated financial return. A great crowd had come to Brunswick to witness the ascent, but many had contented themselves with the view available from outside the Gardens, and from the elevated ground on the South bank of the Yarra River. The anonymous author of *My Note Book* gave a reason for that, writing:

I never supposed (and only wonder that the experienced Mr Coppin should have done so) that any great multitude of full-grown Victorians would be willing to pay five shillings to witness the inflation and ascent of the machine. Had Mr Coppin made no extra charge on this occasion I cannot but think that instead of sustaining a loss, he would have been a considerable gainer.[11]

There was little reaction from the newspapers in Melbourne. The English poet, Richard Horne, now in the colony as a police magistrate and well connected in Melbourne society, wrote to *The Argus*[12] remarking that since 'some gentlemen think that balloons may, one day or other, assist the progress of science in voyages and travels, exploration etc.' the newspaper may wish to reprint an article on aerostation he had written some time before at the request of his friend, Mr Charles Dickens, and which had been published in that gentleman's magazine *Household Words*.[13] It was not an optimistic article, suggesting that dirigibility was beyond the capacity of contemporary science. Horne offered the opinion that it would take a seventy-ton steam engine to provide manoeuvrability for a moderate sized balloon.[14]

Brown had to wait two weeks for his chance to make his ascent:

To *The Argus*, *Age*, and *Herald* Newspapers. Cremorne Gardens, Melbourne, Feb. 15.1858.

Sir, At Mr Coppin's request, permit me to lay before your readers some account of my ascent from Cremorne gardens this evening with Mr Coppin's balloon, the *Australasian*. At 48 minutes past 6 by my watch, I left the ground with a good ascending power, and was much pleased with the picturesque scene at once displayed below me. I first ascended in a north west direction, which was continued until I arrived over the Eastern Hill, at which time I attained my greatest altitude, more than two miles. Here I was seized with a most violent pain in my right ear, which induced me to open the valve in order to descend. The heat of the sun rarefied the gas so greatly, that I was compelled, in order to cause my descent, to open the valve no less than 28 times. From time to time I threw out hand-bills, which I found in my car, the apparent rising or falling of which indicated my ascent or descent. For nearly 20 minutes I was apparently almost directly over Cremorne. At one time I got into a gentle current, which, if it had continued, would have wafted

> me out to the bay; but as there was little daylight remaining, I thought it prudent to descend into the current of air I first met in my ascent; though, had I ascended at an earlier hour, I should most certainly have proceeded over the bay to some distance, and then returned to land by means of the under current. After opening the valve repeatedly, as I have stated, I commenced a very gentle descent, going in an easterly direction over the Collingwood Flat, where I attempted a descent, but, finding I was falling on the houses, I threw out several loose articles I had in my car, as well as all the sand I had taken as ballast. At this time I had reached the current I first encountered, and floating onward at a mean altitude I arrived over a spot which I thought favourable for landing. I drew the valve string, and came down in the most gentle manner possible on the road between Collingwood Stockade and Brunswick, about four miles from Cremorne, after a voyage of 44 minutes. On my descent I was treated in a most brutal manner by the people assembled. Why I know not; but they tore the hair from my head, bruised, crushed, and almost suffocated me, besides damaging the balloon by tugging at and trampling on it. This does not apply to Mr Hugh Peck, of Collingwood, to whom I am under the obligation of returning the balloon to Cremorne in safety, and declining to receive any remuneration for his trouble. Mr Needham, of the gasworks, assisted in extricating me from the savages. In conclusion, I must express my regret that the ascent took place at so late an hour in the day, as if time had permitted, I would have made experiments and observations connected with the southern hemisphere which would have been regarded with interest by the admirers of aeronautics in England. I am Sir, your obedient servant, Charles Henry Brown.

The letter had been ignored by the papers to which it had first been addressed, but *Bells Life in Victoria* obtained a copy and published it, commenting that 'although the actual benefits to mankind arising from balloon voyages have not as yet been proved very important, still there is no doubt that a time will come when aerostation will be reduced to a practical science,' and the writer of the article went on to express 'full sympathy for Mr Brown' and a condemnation of the 'mindless brutality' of those who had attacked him.[15] Six weeks into his Australian adventure, Brown had little to show by way of public response to his endeavour other than this lukewarm endorsement, and the pessimistic offering of Mr Horne.

Notes

1 *The Ballarat Times* 20 January 1858: 3. '*Green's Balloon Extraordinary* will start from the Unicorn Hotel this morning at 11 o'clock. Any person wishing to ascend can enter their names at the Unicorn Hotel before 10am.'
2 This is supposition. But, for whatever reason, Green's Balloon did not ascend from the Robin Hood or the Unicorn Hotels.
3 *The Ballarat Times* 3.
4 The Royal was never a great success, and in 1864 it was sold to the Ballarat Temperance League. Brooke may not have appreciated the irony.
5 In the 'golden age' of industrial pollution in England, the question would have been more appropriately asked with regard to a London ascent.
6 *The Argus* 1 February 1858, sec. Editorial Comment: 4.
7 *The Age* 1 February 1858: 5.
8 *The Age* 2 February 1858: 4.
9 James Magruder, the composer, had probably written the piece to commemorate one of John Wise's ascents. It was published in 1854 by Henry McCaffrey of Baltimore.
10 *Illustrated Melbourne News* 6 February 1858: 6.
11 *My Note Book* (Melbourne: Shaw, Hartnell & Co., March 1858) 470.
12 R.H. Horne, *Argus* 5 February 1858: 6.
13 R.H. Horne, 'Ballooning', *Household Words* 25 October 1851: 83.
14 An unladen Boeing 747 weighs approximately 188 short tons.
15 *Bell's Life in Victoria* 20 February 1858: 2.

CHAPTER 19

Changing Directions

Over the following months all pretence on the part of Coppin – and all hope by Brown – that the balloon ascents were scientifically orientated was abandoned. Attention was focused strictly on making the balloon ascents as attractive as possible to a paying public. Night ascents were made. Fireworks were taken aloft and ignited below the wicker car so that spectators were, at least in the early stages of the flight, able to follow the path of the balloon. Still the Melbourne public remained loath to part with hard cash to watch any spectacle they could view *gratis*. The *Illustrated Journal of Australasia* commented unhelpfully, 'the *Australasian* Balloon has hitherto proved an unsatisfactory enterprise. It has only kept three of several appointments with the public and these are said not to have produced profit for the projectors'.[1]

In other regards the venture was not shaping up as anticipated. Brown and Dean were now flying together, and Dean had no compunction about taking the balloon up when intoxicated. Added to this was the response of the public who appeared to have adopted it as a new game to 'capture' the descending balloon. The object of the sport seemed to be to entice the aeronauts to descend and then try to shake them out of the basket. On a number of occasions the balloon and the aeronauts were manhandled on landing.[2] *Melbourne Punch* gleefully chronicled all these misadventures. By mid-year, Dean was drinking to excess again and while Brown was not prepared to snitch on his colleague, he hoped fervently that Coppin would discover Dean's misconduct and get rid of the man.

The contract Brown had with Coppin specified that he would receive a wage of £3/5/0 a week. Eliza had done occasional work for the aeronauts for which Coppin had allowed her to claim a pound or two. Between the infrequent ascents Brown also worked as a clerk at the Theatre Royal.[3] When winter set in and the conditions were such that flying was impossible he had to fend for himself.[4] It was then that he was contacted by an acquaintance from the Old Dart, William 'Bill' Green. Brown had given up on his earlier attempts to contact Green, and now he was surprised and pleased to be in touch with him. He had no idea what Green had been up to, and Green himself was careful not to mention his earlier aeronautical endeavours at Ballarat. He tentatively suggested a partnership with Brown – he reckoned the pair could make money out of flying. However, Brown was still tied to Coppin by contract for the next season at Cremorne Gardens, so was unable to consider the proposition. To help with finances, he took such work as he could get as a solicitor's clerk, while all the time attempting to engender some sort of debate about the usefulness of ballooning as a means of exploring the interior of the continent. The new 'flying season' was looming when, in September1858, the *The Argus* published one of Brown's letters, extolling the merits of aerial exploration, and suggesting that Australia could be surveyed south to north in as little as eighty hours by balloon.[5] The letter did not spark the discussion Brown had hoped it would. The correspondence column of the *The Argus* was instead taken over by a contretemps between those who were mounting a defence against a charge of being 'willing to make themselves as noxious as possible', by bathing in the Yarra River in the sight of ladies, and those who took the part of the ladies who were distressed by this conduct.[6] Then things took a turn for the better. Coppin decided to make a move.

Coppin had been subject to the sting of insolvency in the past, and was not willing to allow himself to suffer such an indignity again. It was true that in each of the previous instances he had struggled back to liquidity, making one hundred percent restitution to his debtors. But now he had entered politics, being elected to the Richmond Municipal Council, and he clearly had an eye to entering Parliament. In these circumstances insolvency was unthinkable and he needed to forestall another pending disaster with a bold initiative. In an attempt to put the balloon project into the black, Coppin decided to send Brown and Dean to Sydney, there to fly the *Australasian* in the hope that the public would be more generous with their crown, half-crown and shilling donations.

The two aeronauts arrived in Sydney during November 1858. Brown had contrived to take his wife and family with him, and having missed his opportunity to make a mark in Victoria, was excited at the prospect of establishing himself as New South Wales' premier aeronaut. Due to his drinking habits Dean was an unreliable partner, and Eliza was frequently called on by her husband to help and was becoming adept as an aeronautical assistant. She had been promised a flight as soon as it could be arranged. Brown knew from his experience in England that the prospect of a lady going aloft excited the public interest. Dean, for his part, was flushed with anticipation of the high-life on hand for him to experience in Sydney. At the time Coppin's partner, Brooke, was appearing at the Prince of Wales Theatre, Sydney. Brooke did what he could to promote the balloon venture. He took up the ploy of offering a £5 reward for the return to the city of the balloon (and aeronauts) from wherever they should descend. The Governor-General, Sir William Dennison, was inveigled into attending the first ascent on 13 December.[7] The event threatened to get off to an unfortunate start when, caught in a strong wind, the partially inflated balloon collapsed on the Governor's head, but he was rescued without any injury being done, save to his dignity. Brooke anxious to avoid any further mishap dispatched the balloon seven minutes before the scheduled time.[8] The flight was quite spectacular, with a rapid ascent and a favourable current taking Brown and Dean out over the harbour and back.

Brooke decided that the next flight would attract even more notice if his popular co-performer, Harry Edwards, were to be taken aloft. The flight was accordingly arranged and a large crowd gathered, but the press of people was so great that there was little space left around the balloon. One of the struggling spectators stabbed the envelope with his umbrella, and a policeman managed to put another hole in the balloon with his baton. A great deal of gas was escaping. Conditions in the balloon's basket were far from ideal. Edwards was a heavy man, and Dean, as usual, was drunk. Brown thought it best not to make the ascent and left the basket intending to postpone the flight. He turned to warn Edwards of the hazard, but was intercepted by Brooke. 'Don't try to frighten him', he warned Brown.[9] The *Australasian* took off without Brown, and having failed to gain altitude, landed on a housetop, 220 yards from its starting point. Dean fell into the street, injuring his leg. The suspended car swung into a butcher's shop, and Edwards was pitched out, narrowly missing being impaled on a butcher's hook.[10]

Undeterred, Brooke arranged for another ascent. This time a sufficient supply of gas was piped into the Domain and the brave (or foolhardy) Edwards again went aloft with Dean. Edwards had undertaken to regale the audience at the Prince of Wales with an account of his ascent. Safely back on ground the made his way to the Prince of Wales stage and gave a dramatic account of his flight amidst much cheering and applause. The company at the theatre ended the evening's entertainment with a sketch entitled *You Can't See the Balloon*.[11] Matters seemed to be taking a turn for the better.

Brown was approached by an American who made an enticing proposal: he would fund the construction of a new balloon, pay the passage of Brown and his family to California, and top it all with a £100 no-strings-attached fee, provided only that the balloon would bear his name.[12] Brown's enthusiasm for this scheme was doused by Eliza's common-sense observation that having just arrived in Australia, and being in the process of setting up a home and with a new infant, they could not contemplate another move.

Brown and Dean had now established themselves as celebrities. They regularly appeared on the stage of the Prince of Wales and were greeted with great applause. Brown on one occasion went to see a performance by his old acquaintance, John Anderson, The Wizard of the North, who was playing at the Lyceum, but having taken a box seat, his presence caused such a stir that he was obliged to seek refuge in the gods.[13] The Wizard was less than pleased by the interruption of his act.

Brown was once again corresponding with Coxwell in London. His old friend offered sympathy and advice while at the same time seeking what credit he could on the home front for any scientific success that might fall to Brown or Dean. With Brown half a world away, Coxwell felt it safe to lay claim to having been the designer and manufacturer of both of Coppin's balloons, and with having instructed the aeronauts to make meteorological and other scientific observations.[14] At the time Brown did not know of these deceits.

The Sydney-based ascents were providing a reasonable return and flights were scheduled for the new year, 1859. Dean's behaviour, however, was becoming more erratic and he was constantly inebriated. On 17 January a night ascent was scheduled, with Dean insisting that he should pilot the *Australasian*. The flight was to incorporate an aerial firework display, with the fireworks in a rack suspended beneath the balloon. When the time for launching came Dean was too drunk to manage the craft. Brooke was called for and Brown suggested that, Dean being 'unwell', he would substitute. Brooke, unaware as to why there was any sort of fuss over such a matter, gave approval for the change of pilot. Once airborne, Brown discovered to his horror that Dean had failed to put a grapnel on board, and there was no way the balloon's landing could be controlled. Tempted to make an immediate descent into the Domain, Brown found he was unable to do so since the inebriated Dean had loaded such a number of fireworks that they burnt for over ten minutes – normally a display lasted no more than three. Brown was convinced that Dean had set out to murder him![15] When the fireworks were extinguished, Brown attempted a descent, but for want of a grapnel only managed to bounce the balloon across a field. Finally, the wooden and metal frame on which the fireworks had been suspended snagged on a post and rail fence and arrested the balloon's erratic course. Brown was thrown out of the car, but escaped with nothing worse than a bruised shoulder. After the crash-landing the pilotless balloon pulled free from its temporary anchor and floated out to sea. A day later the *Australasian* was found, bedraggled and holed, on the shore near Miller's Point. Brooke decided it was time to call a halt to the venture and for the aeronauts and their retinue to return to Melbourne by the end of January 1859.

Coppin was furious at the curtailment of the Sydney venture. At the end of the previous year he had realised his political ambition and had been elected to the Legislative Council of Victoria as the representative for the South-western Province. He now dissolved his partnership with Brooke. The break-up was acrimonious, and it left Brooke embittered. 'Coppin did not behave well to us,' he reported to a friend, 'But thank heavens I have got the best of it, and have taken possession of my property, the Theatre Royal and the Royal Hotel, valued at £30,000'.[16] Coppin informed Brown that he was not prepared to spend a farthing on repairing the damaged *Australasian*, but he would fund the inflation of the spare balloon and see if this time the public would respond more enthusiastically, or more to the point, whether the venture would now provide a reasonable financial return. When the spare balloon, *Tavistock*, was unpacked it was discovered that the envelope had perished. Make-do repairs were affected, but two attempts to inflate the balloon failed. Despite his initial refusal to repair the *Australasian* Coppin decided on one last fling, and a patched balloon went aloft.

Eliza accompanied her husband on the first flight, but this novelty did not generate any great public reaction. A subsequent night ascent brought in only eight spectators to the Gardens.[17] Coppin had had enough of aeronautical adventuring and now turned his mind firmly to advancing his political career. In a cynical move, designed to convince the voters that he was intent on devoting more time to his political duties, Coppin granted the lease of the Cremorne Gardens to a Mr Grey – his coachman.

Notes

1 *Illustrated Journal of Australasia* April 1858, Vol.4 No xxii ed.: 168.

2 Brown to Editor (*Evening Mail*), *Correspondence*, 30 March 1858: 285.

3 *The Coppin Papers*, Mss. Legers 1858–1859, Box 119/122. La Trobe Australian Mss Collection, State Library of Victoria, Mebourne.

4 Alec Bagot gives the aeronauts' wage as £3/5/0, but the original contract is not extant: vide *Coppin the Great* (Melbourne: Melbourne University Press, 1965) 213.

5 'Balloon Exploration of Australia', *The Argus* 15 September 1858, sec. Supplement: 1.

6 'Bathers in the Yara', *The Argus* 17 September 1858: 5. Thomas Embling, MLC and benefactor of the Botanical Gardens, threatened to take the bathers to court. He found the whole course of conduct particularly disturbing since the swimmers were launching themselves into the Yarra from the landing stage at the Botanical Gardens.

7 'Notices', *Sydney Morning Herald* 8 December 1858: 1.

8 Brown to Green, 16 December 1858, *Correspondence*: 435.

9 Brown to Coxwell, 24 November 1860, *Correspondence*: 318.

10 *Correspondence*: 318.
11 *Sydney Morning Herald* 20 December 1858: 4.
12 Brown to Coxwell, 24 November 1860, *Correspondence*: 322.
13 Brown to Green, 16 December 1858, *Correspondence*: 437.
14 Henry Coxwell, 'Exploration of Australia by Balloon', *The Mechanics Magazine* 28 January 1859: 79.
15 Brown to Coxwell, 24 November 1860, *Correspondence*: 320.
16 William J. Lawrence, *The Life of Gustavus Vaughan Brooke* (Belfast: W & G Baird, 1892) 193.
17 Brown to Coxwell, 24 November 1860, *Correspondence*: 321.

CHAPTER 20

The Dodge

'Mr Coppin, it would appear, if we are honest with each other, that this venture has not succeeded.' Brown was conscious that he was stating the obvious.

Coppin looked at Brown for a moment, and then turned back to the ledger in front of him.

'Will you please free me from my contract?' persisted Brown.

Slowly Coppin closed the ledger, removed his spectacles, and looked up at Brown. 'And what, pray, am I to do with two balloons?'

'If you want to try again at a later date you have Dean. One aeronaut would be sufficient for you, particularly if the ascents don't pay you a reasonable return.'

'Dean be damned!' exploded Coppin. 'I have no confidence in him.'[1] He was about to return to his ledger when a thought struck him. 'Now, Brown, you mentioned an American in Sydney – he indicated an interest in ballooning. Would he purchase my balloons?'

'He was about to return to California,' said Brown.

'I would part with the machines for £100,' said Coppin, as if he had not heard of the incentive offered to Brown.

'Were he here, I would not advise him to buy,' said Brown, and without taking his leave stalked out of the office. Coppin sat for a moment, drumming his fingers on the ledger. This whole business of the balloons was turning Brown into a moody man. Damn him, he thought. Then, with a shrug of his shoulders, he returned his attention to the work in hand. The break-up with Brooke had left finances in a mess.

A day later Brown sought out Bill Green at his place of business in Great Bourke Street, from where he operated as a Land Agent.[2] The establishment was large but shabby. While the many handbills tacked about the walls indicated that Green had potential clients, there was a singular lack of activity about the establishment. The tinkling doorbell brought Green from a small side office. He was a stocky man, clean-shaven, with close-set brown eyes burdening him with a slightly shifty look. For a moment Brown was reminded of his mentor Coxwell's opinion of Green as being 'cunning and selfish', but then Coxwell had not seen Green for many years, and the passage of time brings about many changes. It seemed churlish to think of Green in those terms if he was going to do business with the man – and Brown was in urgent need of someone to join him in his new venture.[3] Despite his earlier exchange with his employer he had made up his mind to purchase Coppin's balloons.

'Why, Brown, I am surprised to see you here,' gushed Green. 'You've come to seek out a neat property for yourself? Emerald Hill is not the most salubrious of locations, you will grant.'

Brown smiled at the superficiality of the greeting. Green was an adept at puffery. 'No, Green, I have already relocated myself. I live now in Cubitt Street.'

'Ah, right adjacent to the Cremorne Gardens!' interjected Green.

'Quite so,' replied Brown. 'But I am here on business of a different ilk entirely.'

'Well, then, let us retire to my office,' said Green, ushering Brown from the main floor. 'You've come to take me up on my earlier offer? You are seeking a partner?' With his hand on Brown's shoulder, he guided him into the office. 'Perhaps we can best deal with each other were you to call me Bill. I may call you Charles, I take it?'

The meeting was not shaping up as Brown had planned. He was the one who came to initiate a negotiation – propose a partnership – and now Green had placed himself in such a way that he appeared to be directing proceedings. Should he now stand on his dignity and insist on the formality of surnames?

'Brown has served well thus far.' He smiled thinly, feeling that in some small way he had regained the advantage. 'I have a proposition,' he said, perching himself on the edge of the wooden chair that Green had drawn up for him. Green gave no appearance of being disturbed by the rejection of his proposal of familiarity. He moved behind a small table and sat down. 'Would you consider joining me in a partnership to purchase and fly Coppin's balloons?' asked Brown.

'But it was I who invited you into a partnership just some five months ago,' protested Green, genuinely surprised.

'The circumstances were entirely different then,' countered Brown. 'I am now in a position where I believe Coppin will release me from my contract, and will sell the balloons. That was not the case when you approached me.' He paused. 'I intend to ask Coppin to sell me the balloons. We will have two machines to fly.' Green looked at him archly, but said nothing. 'Will you join me in this venture?' pressed Brown.

'And what part would you see me playing in this undertaking?' asked Green. 'I have no ready cash available.'

'I believe that Coppin would trust me sufficiently to allow me take the balloons on terms,' replied Brown. 'Would you act as guarantor?'

Green stood up and walked to the door of the office. He opened it and looked into the main hall. There was nobody there. He returned slowly to his seat. 'Guarantor for how much?' he asked cautiously.

'I think Coppin would let me have the machines for £100,' said Brown, sitting back in his chair. 'That was the sum he proposed when he hoped to sell them to an American of my acquaintance. I have got together £50 cash, but am short the of balance.'

'Brooke and Coppin have split as partners, I hear,' observed Green.

'That is so,' confirmed Brown.

'Then I would imagine he has need of whatever cash he can muster. Haggle with him – get the price down.'

'I suspect you know Mr Coppin less well than I do,' said Brown gravely. 'If he has £100 in mind for the balloons. I have already told him they are not worth that amount, but he is unlikely to be shifted from that figure.'

Green grunted and sat for a moment with his chin in his hands. He looked up. 'Half the risk, eh? Then I would expect half the profits.'

'If you would undertake half the management of the balloons,' put in Brown. 'And split the expenses.'

Green stood again. He seemed agitated. 'Yes, yes,' he mused. 'It could be a viable undertaking.' He turned and stood in front of Brown. 'But Coppin has a dislike for me, and may not wish to do business if I am involved.'

'Why is that?' asked Brown.

'A small matter of dispute. An incident in Ballarat some while ago.' Green did not elaborate.[4]

'I am sure matters can be arranged so you have as little as possible to do with Coppin,' Brown assured him.

'Very well,' responded Green, extending his hand. 'I will join you.'

'Excellent!' exclaimed Brown, standing and taking Green's outstretched hand. 'We are partners.'

* * *

Eliza and the children had enjoyed their trip to Sydney, and the children had been thrilled to see their mother go up in their father's balloon. Frederick and Eugene had been most impressed by the reception their father received, and did not understand that the whole undertaking had failed to make a profit. On their return to Melbourne they related fantastical tales of the journey to the other children in Cubitt Street. Annie had acted as nurse to baby Herbert. He had not travelled well, and now back in Melbourne he was running a fever and losing weight. Dr. Gillespie, the local medico, had been called. He prescribed broth and advised that the child be kept well covered, this being the latest treatment calculated to cause the fever to break. On the 25 February, Herbert was discovered dead in his cot. Dr. Gillespie returned, examined the emaciated corpse and pronounced death as being due to marasmus – extreme and chronic malnutrition.[5] The burial took place at Melbourne General Cemetery, and on the same day a distressed Eliza informed her husband that she believed herself to be pregnant.

When he heard of Brown's bereavement, Coppin was duly sympathetic. He was prepared to release Brown from his contract and was easily persuaded to part with the *Australasian* and the *Tavistock* on the payment of £50 cash and receipt of a three-month bill for £50. He required that the bill be guaranteed by a respectable householder and a proper deed of agreement be drawn and signed. But Coppin was moved only so much by Brown's domestic misfortune and his display of sympathy did not extend to paying a solicitor to draft the agreement.

* * *

'I will have the bill of sale and the note attended to, Mr Coppin,' volunteered Brown, since he intended to draw up the document himself.

On the evening of Thursday, 31 March 1859, Brown laid the documents before Coppin. William Green, of Victoria Cottages, Carlton Gardens, was cited as co-signatory of the promissory note for £50. Brown contrived to position Green in a dark corner of Coppin's office and he hoped that the formal presentation of Bill Green as William Green, and the use of his residential address rather than his business, would confuse Coppin as to the actual identity of this third party. As soon as the promissory note was signed, Brown hustled his new partner from the office and then completed the agreement of sale for the balloons. The canny Coppin had insisted that a restraint of trade be incorporated in the document. Brown bound himself 'until the expiration of the year 1860 not to make balloon ascents (or allow balloons to ascend) for any person in Melbourne but the said George Coppin'.[6] The new agreement was duly signed, while the original contract made between Brown and Coppin in London was cancelled.

'I wish you success with this venture, Brown,' said Coppin graciously. Brown nodded as they shook hands. 'Bye the bye,' remarked Coppin, glancing at the document on his desk. 'This Green chap; not a kin of Bill Green, I take it.'

'That was Mr William Green, the land agent,' responded Brown.

'Good God, man!' exploded Coppin. 'I want no business with that rogue.'

'I do not imagine you need worry on that account,' replied Brown calmly. 'The promissory note will be redeemed on the due date, and we need not worry Mr Green about the matter.'

'Worry indeed!' snorted Coppin. 'Had I known Green were involved I would not have let the balloons go.'

'I am sorry you have taken it so,' said Brown, 'but I believe you will not in any way regret our agreement.'

Coppin was far from being mollified. 'I hope not, Brown. I sincerely hope not,' he remarked, and pointed to the door. 'I bid you goodnight.'

Brown went to meet Green at Cubitt Street as they had arranged. He wondered what had occurred between Green and Coppin to leave the latter so upset. Whatever it was, he decided, was a matter of the past and best left there. There were exciting times on offer in the immediate future.

Notes

1 The conversation, to this point, is detailed in Brown's letter to Coxwell. 24 November 1860, *Correspondence*: 322.
2 *The Argus* 17 September 1853: 6. also 'Insolvency', *VPRS 75/P-3*, 1861.
3 Coxwell to Brown, 1 November 1858, *Correspondence*: 295.
4 The 'incident' was Green's attempt to pre-empt Coppin's launch of the *Australasian* by taking his own balloon aloft at Ballarat during January, 1858. Terence O'Neill-FitzSimons, 'A Balloon on the Ballarat: Green's Balloon Extraordinary', *Victorian Historical Journal* Vol. 71. No. 1. (March 2000): 19.
5 Registration of Death, Certificate 959/1859. *Victorian Registry of Births, Deaths and Marriages.*
6 *The Coppin Papers*, Mss 8827, Box 12/18. La Trobe Australian Mss. Collection. State Library of Victoria, Melbourne.

CHAPTER 21

The Reprise

Within a few days of concluding the purchase of the balloons, Brown decided to take the *Australasian* to Sydney and to use it to initiate the new venture.[1] With the death of his infant son just behind them and Eliza newly pregnant, he was not prepared for his family to join him on this trip. Green was quite content to have his new partner to look after all the troublesome and pernickety details of preparing the balloon, netting, car and appurtenances. The *Tavistock*, repaired in a makeshift way and fitted with a new car, was left in Melbourne in the custody of a friend, a Mr Bellin, who had loaned Brown £30 to help secure the deal with Coppin.

Already a celebrity in Sydney, Brown assured his partner they would meet with a good reception when they arrived. But he did not have the same connections that Brooke had able to call on, and there was no certainty that he could secure a good venue for the proposed ascent. When last in Sydney he had struck up an acquaintance with Abraham Polack, an auctioneer with a colourful background.

In his younger days Polack had gained a certain notoriety by taking his horsewhip to a local who had 'mentioned his name to a widow lady in a disrespectful manner'.[2] When Brown had been in Sydney Polack had expressed to him his interest in aerostation. He was running two steamers on the Hunter River and the potential of aerial transport intrigued him. He had a large vacant plot between George and Pitt Streets at the rear of the Post Office. This was a central and accessible location and spacious enough to accommodate a balloon. Polack offered Brown the use of the sit. It now remained to obtain an adequate gas supply.

Green, meantime, had been touting for business. It had been decided by the partners that each ascent must be made to pay for itself. Green had undertaken to look after the monetary side of the business and it was agreed that a sensible way to meet their financial objective was to carry fare-paying passengers on the flight. It was at this time that the first rift in the relationship between Brown and Green appeared. Much to Brown's annoyance, he received reports that Green had been boasting that the citizens of Sydney hadn't seen anything yet in the way of decent aerial ascents, but he would shortly remedy that with a display of aeronautic virtuosity. Brown was slightly mollified when Green produced their first fare-paying passenger. Curiously, it was Alfred William Wardell, son of the proprietor of the Vauxhall Gardens in London, where Coxwell had previously ascended, and from where young Wardell had undoubtedly gone aloft. Why he should be prepared to lay out good money for something less than a novelty was not clear unless it was 'for old time's sake', since the Vauxhall Gardens were about to close and Wardell senior had been thrown into bankruptcy.[3] Green reported that he had negotiated with the young Wardell for a fare of £20 to be paid on the day of the ascent. With the prospect of the first flight actually paying for itself Brown agreed, when pressed, that Green could take up the *Australasian*.

Monday 18 April 1859 was nominated as the day for the ascent and handbills and posters were put about announcing the event. Under instructions from Brown, the Gas Light Company had rigged a pipe from the town mains through to Polack's vacant plot. Brown planned to make the affair a gala event. Conscious that Green had a greater flare for dealing with the public, he persuaded Charles Poole, at the *Prince of Wales* theatre, to allow Green to appear after the ascent and relate the particulars of the trip to the audience. This had proved popular with the public when Henry Edwards had done the same thing just some months before during Brown's last visit. Poole agreed. All was ready for the flight.

The announcements of the ascent had excited considerable interest. The weather on the day was perfect. Brown released a number of pilot balloons, and with little or no wind they rose to a great height. The *Australasian* was scheduled to take to the air at five o'clock and by that time a vast concourse of spectators had assembled in the neighbourhood. There were reserved seats close to the balloon, but most of the spectators preferred to remain in Pitt Street or else perch on the roofs of adjacent houses, while many had stationed themselves at the windows in the rear of the Post Office.[4] Brown was satisfied that the receipts from the small number of paying spectators, when combined with Wardell's £20, would be sufficient to cover expenses. He had every hope that Green would be able to generate more paying customers for future flights. He would have the opportunity to spruik those events from the stage at the *Prince of Wales*. At five minutes past five, with Wardell safely on board and the balloon fully inflated, Green gave the signal for those holding the tethering ropes to let go, and the *Australasian* rose swiftly over Hyde Park.

Brown began to time the flight. The balloon had been in flight for eight minutes and it seemed to be drifting further towards Sydney Cove. It then moved back inland, tracking towards the south. The *Australasian* looked set for a long successful flight when suddenly it commenced a rapid descent in the direction of Brickfield Hill at the end of George Street. Brown and many hundreds of the excited spectators quickly followed.[5] The machine was descending into the yard of the *Woolpack Inn*, where there were already a considerable number of people milling around. A line was dropped and the balloon was being hauled into the yard when Green thought better of the manoeuvre and opted to reascend and land in a nearby paddock. By now a startled Wardell had added his voice to Green's and both men shouted for those on the ground to release the rope and allow them to re-ascend. In the confusion and excitement of the moment, those on the ground did not hear the instructions, or out of devilment chose to ignore them. The crowd held onto the ropes and continued to haul in the balloon. By now the area was packed with spectators. Those who could not find their way into the yard perched themselves atop sheds and fences while others crushed into the laneways as they watched the giant machine lumbering about.[6] To add to the confusion, there was livestock in amongst the crowd.

The owner of the *Woolpack Inn* had four or five bullocks corralled in his yard – a convenient source of supply of fresh steak for his hostelry. Unfortunately, as the balloon swayed and billowed, one of the animals took fright and in panic speared the envelope with its horns. Gas rushed from a large rent and young Wardell, overcome by the fumes, was lifted out of the basket in a state of unconsciousness and carried into a shed. Now lightened by the loss of its passenger, the *Australasian* became more buoyant and Green spotted the opportunity of making an escape from a situation which was becoming totally out of control. He again shouted for the ropes to be released, believing that he could float the balloon out of the yard and land further to the south in open country, but the people holding the ropes were still loath to release their prize. By now the police were on the scene. Captain M'Lerie and a number of constables forced their way into the yard where the ripped *Australasian* was still held fast and venting gas.

Brown had joined the crowd in their pursuit of the descending balloon, but when he got to the Inn's yard he was unable to move any further, being held back by the heaving, shouting throng. In the midst of this heaving crowd the terrified cattle remained trapped. By now Captain M'Lerie had been joined by members of the fire brigade, under the direction of their Superintendent, J.T. Brown, but their combined efforts failed to bring about a restoration of order. On the edge of the mob Brown smelt the escaping gas and noticed the rip in the balloon. Then to his horror he spotted an open forge nearby. He feared that the gas would 'get some fire – either from the blacksmith's shop or houses'.[7] The danger did not lie however in the smithy's forge or the domestic gaseliers. One of the locals had seated himself atop an adjacent shed to view the spectacle, and comfortable in his shirt sleeves with his cabbage-tree hat tucked under his arm, he lit his pipe and casually tossed aside the still flaming match.[8] (The report of the incident in a Melbourne newspaper refers to this unknown miscreant, without exaggeration, as 'an ignoramus'.[9]) A terrific explosion instantly ensued. According to one account, 'Vivid lines and sheets of flame traversed the yard in all directions like flying fiery serpents'.[10] The gates that led into the yard were now packed with people attempting to escape the conflagration, while many who had been trying to push their way in could not move aside and contributed to the crush. Eventually some of the surrounding fences gave way and the crowd spilled out into Durand's Alley and George Street, with women and children being trampled underfoot and injured. One young woman had her face crushed by a boot when she was pushed to the ground. Meanwhile Green, in an attempt to speed the deflation of the fiery balloon, further cut the envelope but only succeeded in creating a fireball. Enveloped in flames, the *Australasian* rose majestically into the air for a moment and then fell back to earth, reduced to ashes and its car a smouldering bundle of wicker-work. Green was now trapped in the press of people and could not get away. Fortunately no one was killed though many were injured, with about a dozen seriously so.[11]

Brown had managed to free himself from the crush, and now stood a little away from the mêlée as people fled with their hair on fire and their faces horribly disfigured.[12] Memories of the ghastly incident at Bradford involving his *Monster Balloon* came flooding back. As he looked at the scene of devastation some of his friends, who had been frantically searching for him, grabbed him and bustled him into a cab. They believed that once the panic abated there was every likelihood the crowd would turn nasty and be out for Brown's blood. He was deposited at Fischer's Hotel in Castlereagh Street, a safe remove from the scene of the air wreck. Green, who was still at Brickfield Hill, decided that his best course of action was to excite what sympathy he could. Claiming his ribs were broken, he had himself carried to the residence of a local surgeon. When he met up with Brown a few hours later there was an angry exchange between the partners. Brown dismissed Green's claim to being injured as a complete sham and accused him of having mismanaged the descent to such an extent that he must take responsibility for the whole disaster.[13] Green limped away in a huff. He swiftly had his revenge. Meeting a reporter from the *Sydney Morning Herald*, he gave him an interview, telling how, having invested 'all the money he had in the world, in the purchase of the balloon', he was now faced with absolute ruin. When the newspaper carried this piece the following day, Brown was irritated. This irritation changed to rage when Green, attempting to make his peace with his partner, confessed that he had not been entirely truthful about the fee paid by Wardell for the flight – he had only given £3 for the privilege of flying in the *Australasian*. The whole venture was now seriously in the red.

The reports of the balloon's wreck carried in *Bells Life in Sydney & Sporting Review* were not sympathetic. The newspaper laid the blame for the calamity at Green's door. 'Had Mr. Green acted with his usual precaution', the article opined, 'he would have effected a descent beyond the city. While we sympathise with him in his misfortune, we cannot but in a measure ascribe it to an error of judgement'.[14] The situation was not entirely lost, as Brown still had his supporters. Poole and Edwards of the *Prince of Wales* joined with Polack in organising a subscription for the two aeronauts, and

Poole further undertook to run a benefit night at the theatre. Brown for his part opted to go on the attack and wrote a letter to the Editor of the friendly *Sydney Morning Herald*. It was published on 22 April:

> Sir – We beg, most respectfully, to tender our grateful thanks for the sympathy with us, as expressed in your columns of Thursday last. But whilst we deeply lament the ruinous consequences to ourselves, as regards a heavy pecuniary loss, and the personal injuries sustained by spectators, we will request to express publicly, through your column, the deep sense which we entertain of the great kindness we have received since our arrival in Sydney, both before and after the ascension. A committee of gentlemen of influence in this city has been formed to consider the best means of raising subscriptions to assist us. To them, and to many other kind friends, we desire to offer the expression of our sincere gratitude. We cannot, however, conclude this letter without adverting, in language more in sorrow than in anger, to the illiberal spirit under which thousands of citizens and their families witnessed the inflation and ascent from the roofs, walls, passages and streets *outside* the ground, refraining from paying the very small fee for admission at the gate, and the reserved seats for ladies.

This Parthian shot of Brown's was probably ill advised, abating whatever public sympathy there may have been for the aeronauts. The subscription and benefit together raised only £60, and this had to be paid to various creditors before Brown departed from Sydney.[15] It was a crestfallen duo that returned to Melbourne five weeks after their departure. Brown was now left with outstanding debts and an un-airworthy balloon, the *Tavistock*.

Notes

1 Brown to Coxwell, 24 November 1860, *Correspondence*: 323.
2 *Australian* 16 March 1838.
3 William Kent, 'Vauxhall Gardens', *An Encyclopaedia of London* (London: Dent & Son, 1932): 692.
4 *Sydney Morning Herald* 19 April 1859: 5.
5 Argus Correspondent, *The Argus* 20 April 1860, sec. Telegraphic Despatches: 4.

6 'The Balloon Ascent on Monday', *Bell's Life in Sydney & Sporting Review* 23 December 1859: 4.
7 Brown to Coxwell, 24 November 1860, *Correspondence*: 324.
8 'Balloon Ascent and Terrible Explosion', *Sydney Morning Herald* 19 April 1859: 5.
9 *Bell's Life in Victoria* 30 April 1859: 2.
10 *Sydney Morning Herald* 5.
11 *The Argus* 20 April 1859: 4.
12 Brown to Coxwell, 24 November 1859, *Correspondence*: 323.
13 *Correspondence* 324.
14 *Bell's Life* 4. It is curious that the reporter professes to have some insight into Green's level of aeronautical skills, since he had never been aloft in Sydney. He may have read of Green's English exploits or even seen him back in Britain. Or, could it be that the correspondent had witnessed an ascent by Green in Ballarat or even Melbourne?
15 It cost £60 to fill the *Australasian* with gas. *Correspondence*: 319.

CHAPTER 22

Hard Times

It was towards the end of April 1859 that Brown approached Coppin, seeking employment. There was no disguising the extent of his predicament; the Melbourne papers had carried reports of the demise of the *Australasian*. Coppin, although a hard-headed business man, had a soft spot for Brown and bore no grudge against him for the subterfuge he had used to complete the contract for the sale of the balloons. He was in the process of disposing of the leases on his theatres – properties not as easily dealt with as the Cremorne Gardens. Brooke was claiming the Theatre Royal as his share from the dissolved partnership. Coppin still had the grandly named Royal Olympic Theatre on his hands, and he was prepared to take on Brown as money-taker and box-office keeper.[1] He was mindful too that there was only two months to run before the outstanding promissory note for the purchase of the balloons became due – best, Coppin thought, that Brown should have some cash to hand; even if it were, effectively, money from his own purse!

Grateful as he was for the employment, Brown needed something more permanent, and while he worked effectively in the box-office he treated the position as a stop-gap. There was the constant pressure of a pregnant wife and three children to provide for at home and Brown needed a steady job, with better pay that Coppin was prepared to allow. A month later he had secured a position as a clerk with a firm of solicitors in Collins Street. He had not, however, given up on the idea of making a success of aerostation.

Suddenly, the prospect of another balloon ascent presented itself. For a number of years the central Victorian goldfields had been an attractive venue where theatrical performers and entertainers could display their skills.[2] It was not that the places of amusement themselves were necessarily comfortable or well appointed – no less a personage than Madame Lola

Montez had, in the past to contend with collapsing sets and burning scenery during her goldfields visit – but the rewards were great, with enthused miners likely to throw gold nuggets at the artists.[3] During August, Brown received a communication from the gold-mining town of Bendigo. Would he be interested in making two balloon ascents at that place? A fee of £100 was proposed for the engagement. If he were interested a Mr Joseph Field would visit and finalise the arrangements. The restraint in trade agreement Brown had signed for Coppin was limited to Melbourne, so there would be no problem in that regard. The *Tavistock*, though in poor condition, still offered the prospect of going aloft again. After its purchase Brown had deposited it with a friend, and instructed that it be kept well hidden from Coppin. This was an easy enough task, as the balloon and the netting folded down nicely into the wicker car and could be tucked away inconspicuously in the corner of a shed. The move had been a hedge against anything going wrong with the Sydney venture, and calculated to make its repossession difficult. Brown was easy in his conscience over the subterfuge.

Brown was very interested in flying at Bendigo. So too was Green, who got to hear of the venture after Brown had been to check on the hidden *Tavistock*. Green had colluded in the concealment of the balloon, and over the months had kept a wary eye on the premises where the craft nestled. Now it would be an easy matter for him to keep check on the Cobb & Co. arrivals from Bendigo and intercept Field on his arrival in Melbourne. And intercept him he did.

Green grandly introduced himself to Field as the sole proprietor of the *Tavistock*, and puffed his exploits in Sydney. It was unfortunate for him that Field had conducted some investigation concerning local aeronauts, and he knew of the opinion held by the *Sydney Morning Herald* (as copied in *The Age*) of Green's inept ascent from the Domain. Field caught the next Cobb back to Bendigo. When Brown got to hear of the incident, he was furious. He decided to do something that had been at the back of his mind ever since his return from Sydney.

* * *

Perched behind his high writing desk in the corner of the small office, Brown had considerable privacy. He was situated just inside the main door, and his space served as a reception area for clients. His experience as a clerk in the Post Office at Leeds stood him in good stead and he quickly established a reputation as a solid, reliable worker. The senior clerk tended to leave him to work at his own pace, away from the juniors who were cloistered in a large inner office.

Laying aside his pen, Brown carefully placed a sheet of blotting paper atop the letter he had been writing. He watched as slight blue veining appeared on the white sheet. He felt that the missive was both pertinent and pithy.

> To Mr William Green, 51 Stephen Street, Melbourne.
>
> I hereby give you Notice that the Partnership hitherto existing between us, as Aeronauts, is this day dissolved by me.
>
> Charles Henry Brown, May 11th 1860.

He would have the messenger-boy deliver it that morning, and that would be the end of the matter. He glanced once again at the letter, nodded to himself, and, well satisfied, folded the document and sealed it.

That afternoon Green arrived at the office, angry and drunk. 'What in damnation are you at, Brown?' he demanded.

Brown glanced anxiously towards the inner office door. 'Why Green, what brings you here?' he asked artlessly.

'What in hell's name do you think brings me here?' shouted Green, producing the crumpled notice of dissolution from his coat pocket and waving it. 'This, this …' He was spluttering and spitting, and so flushed in the face that Brown feared he might collapse or be taken by a convulsion. 'This rubbishing twaddle!' he finally managed. 'You cannot do this.'

Coolness is what is called for, decided Brown. Having stepped from behind his desk in anticipation of Green falling, Brown returned to his stool with a great display of nonchalance and picked up his pen. 'I am, in fact, glad to see you,' he said in what he hoped was a conciliatory tone of voice. 'I have something of importance to communicate.'

'Have you not bloody well communicated it already, you rogue,' snarled Green. He had dropped his voice, but his manner was no less threatening for that.

'I must ask you to moderate your language,' said Brown pompously.

'Bugger moderation,' countered Green. 'You've left me owing Coppin fifty bloody pounds. Where am I to find that sort of money?' he finished pathetically.

'I am prepared to help,' offered Brown. 'Technically I am a co-respondent.'

Green had by now gained his second wind. 'What!' he exploded. 'You'll help discharge the debt? I doubt that. You hadn't the needful in the first instance, I doubt you have it now, Mister Clerk,' he sneered. He paced about the small office, every now and then thrusting the notice of dissolution under Brown's nose and grunting, 'Huh!'

'If you will but be still for a moment, I have something of importance to relate.'

'Something of importance? Something of importance?' Green was now shouting again. 'You have ruined me. Is that a matter of sufficient importance?'

'And you have interfered with my business,' shot back Brown. He did not raise his voice.

'Are you talking about the *Tavistock*?' roared Green. 'That machine is as much mine as it is yours, you blatherskite.'

'If you choose to continue in this abusive way, I will not tell you what I intended.' Brown was steeled in his resolve to remain calm in the face of all provocations. He picked up his pen and began to write, but he carefully chose a piece of scrap paper, and what he laid down was nothing more than a scribble.

'God blast you, Brown, you loquacious cock-fart! You are talking like an old woman, and acting like a muff.'

In spite of himself, Brown made one last effort to calm the situation and bring about a resolution. 'Coppin, I do assure you, will not be in a position to take advantage of either you or me, if you will but listen to what I have to say.'

'Listen to you!' snorted Green. 'I would just as soon jump in the Yarra Yarra. Coppin will take me to law over this.'

'Well, I can assure you I am in as much peril from the law as you, but if you listen to me, you will find there is no need for alarm,' said Brown, soothingly, but Green flared up again.

'You are talking through your arse. Here, take your damned notice and stick that in the same place.'[4]

Shocked, Brown sat open-mouthed at his desk as Green threw the crumpled paper on the desk and slammed out of the office. After a moment, Brown picked up the letter and smoothed it out. He was satisfied he had written precisely what he wanted to convey, but was flabbergasted by the response it had provoked. After some moments he recovered, and muttered to himself. 'It must have been the drink.'[5]

* * *

Despite all his setbacks, Brown had in no way given up the idea of taking to the air again. He regularly called in at the Mechanics' Institute on Collins Street to read up on the latest articles relating to aerostation. On one such visit, Brown found the latest copy of the *Mechanics' Magazine*, just arrived from England. It boasted an article by Coxwell entitled, 'The Exploration of Australia by Balloon.'[6] Brown read it with a mixture of excitement and irritation. Coxwell was identified in the introduction to the article as the 'designer and constructor' of the *Australasian*.[7] Brown was prepared to ascribe that claim to an ill-informed editor. He was gratified to find himself billed as the aeronaut in charge of the colonial venture, with Dean marked off as his assistant, though he was irritated by the inference that Coxwell was somehow overseeing the 'required scientific observations'. While at the Mechanics' an acquaintance drew his attention to a piece in the *The Era*, a London publication directed towards members of the theatrical profession. Headed 'Professor Anderson in Sydney', it was a three-quarter page spread, written by Brown's erstwhile travelling companion, The Wizard of the North.[8] While it contained a deal of puffery regarding Anderson's exploits in the colonies and the large houses he was drawing, the article concluded with details of Coppin's venture into the field of aeronautics. Anderson informed the reader that the undertaking had been less than successful. He told of the *Australasian*'s visit to Sydney in December 1858, and the ascents made at the beginning of 1859. In particular Anderson had chronicled Brown's flight, which had terminated in the temporary loss and damage of the *Australasian*. 'This was the culmination of aerial disaster,' wrote Anderson. 'Not only had the ascents been more or less failures, and not only had the public since the first attempt been chary of paying their admissions, but now the balloon was lost into the bargain. It was subsequently discovered a wreck on the sea-shore'. Brown had to allow that there was more than a grain of truth in what Anderson had written, but he suspected there was a greater element of jealousy involved. The great Wizard had not taken kindly to being upstaged in Sydney. Anderson concluded his piece by promising his readers further information from the colonies supplied by their own 'special commissioner'. Brown was thankful that the Wizard had departed by the time he and Green had returned to Sydney in April, and could therefore not have witnessed the *Australasian*'s fiery end. But he was mindful of the need to protect his reputation, and very conscious of how the public lapped up news of ineptitude. He quickly put pen to paper. Brown wrote to *The Era*, claiming that Anderson's observations were inaccurate and his comments made him appear 'a bungling aeronaut'. He then quoted some of the more favourable fragments from the *Sydney Morning Herald*'s various reports of the *Australasian*'s ascents.

On the heels of all his disappointments, Brown had commenced a correspondence with his brother, Edwin. Edwin was prepared to be sympathetic – a response Brown knew not to expect from his father, still alive and residing in Leeds, and with whom he never communicated directly. Edwin had been befriended by Coxwell and had assisted him on occasions with the ground management of his balloon, but he was cautious about the whole aerostatic business and encouraged his brother to keep to steady employment. 'It would have been a sad affair for thy family if thou had lost thy life in the late affair', he wrote with regard to the Sydney crash-landing.[9] He also cautioned against continuing in any sort of partnership with Green. This he had picked up from Coxwell, who considered Green untrustworthy, and had warned 'partnerships are the worst ships you can sail in'.[10] Brown was quick to assure his brother that he had dealt with that matter, and in turn was encouraged to be told by him how Coxwell, in the latest edition of his *Aerostatic Magazine*, had defended the Sydney ascents against Anderson's criticism.

Eliza's fifth child was born in November – a son named Vincent. The infant's arrival softened to a degree the sorrow of Herbert's premature death. Brown continued diligently in his clerical post, yet took time to write to *The Age* extolling the merits of the drag and sail system as a means of imparting some degree of dirigibility to balloons, and again promoted the idea of a balloon exploration of the interior of Australia. At the same time the debate on the merits and de-merits of bathing in the Yarra River, now in its third month, carried on in the same letter columns. Gentlewomen taking their morning perambulation continued to be shocked by the sight of men, clad only in their undergarments, swimming in the Yarra.

The particular restraint in trade into which Brown had entered with Coppin had over a year to run and the promissory note held by Coppin with respect to the sale of the *Australasian* and the *Tavistock* had fallen due twelve months earlier. Not only was there money owing – which Brown did not have, and for which the furious Green had stood guarantor – but the agreement also provided that the machines be forfeit to Coppin. That provision was of little consequence in regard to the destroyed *Australasian*, of which there remained only a charred car, and that still in Sydney. The *Tavistock* was safely hidden away.

Coppin knew that Green's business as a land agent was doing passably well, and there might be a hope of recovering the £50 from that source, and Brown was in gainful employment; perhaps he could seek the issue of a garnishee; and then there was always the *Tavistock*. He considered his options.

Even as he did so, Brown took steps to protect himself – and his income. On 5 July 1860 he sought a Certificate of Insolvency from the Supreme Court. One of the solicitors at the firm for which he worked, Mr Bayne, had offered to lay the process before the Court, provided that Brown himself drew the required papers.[11] In the end the process had left him out of pocket only £3. There would have to be a creditors' meeting, after the appropriate notice had been placed in daily newspapers, and a scheme of payment of debts devised. Coppin was the only creditor of consequence and Brown calculated that he was unlikely to spot the notice, or if he did see it, he doubted he would take the time to attend the meeting and wait around while the Commissioner of Insolvents went through his tedious rigmarole – a tiresome process with which he was all too familiar. Five weeks after his application, Brown met Green in the vestibule of the Court. Green had, independently, also decided that the best way to protect himself against any claim by Coppin was to seek bankruptcy. He had not realised just how complicated the process could be for a layman to undertake unassisted.[12] He was contrite for earlier drunken behaviour. Brown accepted a mumbled apology without comment.

Shortly after this meeting, an opportunity presented itself to Brown to resolve another enmity. Dean had been working at Donnybrook, about twenty miles north of Melbourne.[13] Returning from work one evening in a drunken state, he had fallen from the mail-coach and broken a leg. On hearing this news Brown went to visit him in hospital. A chastened and sore Dean was genuinely pleased to see Brown, and offered to provide his labour – free of charge – should it be needed in the event of another balloon ascent being undertaken. Surprisingly, not long after this meeting a request came for Brown to make an ascent. It came from Coppin.

Notes

1 W. Pitt and R. Younge were to take the lease on the Royal Olympic Theatre. Melbournians referred to the establishment, situated on Stephen Street (now Exhibition Street) and the junction of Lonsdale Street, as 'The Iron Pot'.

2 Charles Waller, *Magical Nights at the Theatre*, ed. Gerald Taylor (Melbourne: Gerald Taylor Productions, 1980) 27.

3 The incident of the fiery scenery occurred at Bendigo during a performance, April 1857. Montez had also been engaged by Coppin to play at his Theatre Royal in Melbourne. *Lola Montez-The Tragic Story of a Liberated Woman* (Melbourne: Heritage Publications, 1973) 73.

4 Green's choice language was compiled with the assistance of Eric Partridge's, *A Dictionary of Slang and Unconventional English*, ed. Paul Beal, 8th ed. (London: Routledge, 1984).

5 This confrontation is mentioned by Brown in a letter to Coxwell, written more than a year after the event. He particularly makes mention of Green's 'vile' language. 24 November 1860, *Correspondence*: 325.

6 Henry Coxwell, 'Exploration of Australia by Balloon,' *Mechanics' Magazine* 28 January 1859: 79.

7 Coxwell perpetuated his (false) claim to being the builder of the *Australasian* when he repeated it in his autobiography, *My Life and Balloon Experiences* (London: W H Allen & Co., 1889) 33.

8 John Anderson, 'Professor Anderson in Sydney', *The Era* 17 April 1859: 6.

9 Edwin Brown to Brown, 10 August 1859, *Correspondence*: 305. The brothers in their letters made free use of the dialectic 'thee', 'thou' and 'thy'.

10 Coxwell to Brown, 25 March 1869, *Correspondence*: 310.

11 *Insolvency Records*, Supreme Court of Victoria. *Certificate of Discharge*, 8 November 1860, VPRS 75/P/3 Public Records Office, Melbourne.

12 Green did not manage to get his application before the Court until January of 1861. *Insolvency Records*, Supreme Court of Victoria. *Application for Sequestration*, 11 January 1861.

13 Renamed *Kalkallo* when another 'Donnybrook' was established on the new railway line in 1872.

CHAPTER 23

A New Balloon

The *Tavistock* was on the market! Alarmed by the secrecy and the surreptitious checking of the balloon, Bellin, its custodian – to whom Brown was already indebted in the sum of £30 – opted to exercise his lien over the property and offered the balloon to Coppin. Even though the legend over the entrance at Cremorne Gardens announced Mr Grey as the lessee, Coppin still acted as proprietor behind the scenes. He was prepared to give aerostation one more fling. He wrote politely to Brown:

> Cremorne, Sept. 22.1860
>
> Sir,
>
> The balloon 'Tavistock' has been offered to me by Mr Bellin. If you will undertake to repair it, and make four ascents during the next season at a reasonable price, I will purchase it.
>
> Yours respectfully,
> Geo. Coppin

Believing he now had the advantage, notwithstanding Bellin's 'betrayal', Brown countered with some rather elaborate 'proposals' of his own.[1] For a start, he would not fly in the *Tavistock*. He would instead build a new balloon, half the size of the *Australasian* and capable of lifting two people. In addition this machine was to be named, *Brown's Balloon*. Further, Coppin was to employ Brown for twenty weeks at a wage of £5 a week, pay for the inflation and recovery of the balloon, and contribute £40 to the cost of construction. At the end of the engagement the balloon would pass to Brown as his property. The second scheme was similar, except that Brown would make four free ascents only, and then take ownership of the balloon.

Surprisingly, Coppin was prepared to consider the latter proposal, but he countered by offering £5 for each of the four ascents, and warned he could get someone else to do the job for that amount and still have a balloon at the end of the bargain. In a huff, Brown broke off negotiations.

Green, in the meantime, had somehow come to hear the news that a balloon ascent was in prospect, and promptly offered his services to Coppin. He announced he would undertake the flights at a paltry £2 a time. Nothing had occurred over the months to abate the ill-will Coppin felt toward Green, and he promptly turned down his offer to act as aeronaut.[2] What Brown had discovered during all of this was the asking price for the *Tavistock* – £20. He believed he could raise the £30 to cover his outstanding indebtedness to Bellin and once he had paid this he could reclaim the balloon. It would take a considerable amount of work to repair the craft, and even then it might not be fully airworthy, in which case he would destroy it. Either way he would have the satisfaction of putting it across Coppin, and Green.

But he was still an undischarged bankrupt. He would need to obtain his certificate of discharge before he could incur any further debt without risking a legal penalty. He would need to borrow if he were to build a new machine or repair the *Tavistock*. He set about drawing up the necessary forms of application, but by the time he had submitted them to the Supreme Court and had them processed, Coppin had made his move. He had gone ahead and purchased the *Tavistock* and moved it to the Cremorne Gardens. Thus frustrated, Brown decided that any hope of making his mark as an aeronaut probably lay outside Victoria. He refined his plan. Coppin would, without doubt, attempt to make matters as difficult as possible if the flights were to be made in Melbourne. There was the restraint of trade agreement with Coppin, applicable to Melbourne, binding him until January 1861. Anyway, Green had already queered the pitch in the goldfields. New South Wales was out of the question after the fiery demise of the *Australasian* in that colony. Tasmania seemed to offer the best prospect. Brown was still an avid reader of magazines and newspapers, searching and researching for items on aeronautica. He concluded, erroneously, that Tasmania had never seen a balloon, let alone witnessed an ascent.[3] There was still, however, the question of funding for such a venture. He decided on Hobart Town as the best venue, and placed an advertisement in the local newspaper:

> *BALLOON.* Mr C.H. Brown, Aeronaut, Conductor of the First Balloon Ascent in Victoria and New South Wales, having concluded his engagement with the Hon. George Coppin, will be happy to treat with any gentleman, or a number of gentlemen, for one or more ascents at Hobart Town. To those gentlemen who may be desirous of introducing a Balloon to Tasmania, Mr Brown would suggest the formation of a committee to superintend the matter, and that members of such committee undertake the disposal of tickets of admission to witness the spectacle to an amount sufficient to cover expenses. Letters addressed to C.H. Brown, Aeronaut, York Place, Emerald Hill, Melbourne, will meet with prompt attention.[4]

The plan was for the balloon to be displayed, inflated, in an appropriate hall to whet the locals' appetite and then an ascent would be made. Brown would need an assistant, and using Green in this role was out of the question. He decided he could trust the 'reformed' Dean. The whole project foundered when there was no response to the advertisement.

Obstinately, and not a little insensitively, Brown made another approach to Coppin. He was desperate to secure the position as the colonies' premier aeronaut. He wrote offering to make an ascent during the Christmas holiday period – but he had not learnt his lesson. He asked £20 for one ascent, and stated he would be prepared to go aloft twice for £35.[5] Coppin was bemused by this display of persistency, and declined to engage Brown on those terms.

While the year ended without Brown having returned to the air, 1861 would see the restraint of trade that Coppin had imposed on him lapse, and he would be free to fly in Melbourne. It was now a question of finding a sponsor or a balloon. He began the New Year by writing to *The Argus* extolling the benefits to be obtained by explorers should they choose to use Montgolfier balloons to assist them in their adventure, this was in response to a letter from 'E.B.S' suggesting a gas balloon would be an asset on exploratory trips.[6] Brown's letter was not published (though the debate over bathing in the Yarra still continued in the letter columns), but he did not given up on the idea of a hot-air balloon as an aid to exploration. He had spoken to Dr Ludwig Becker about this before the naturalist had embarked on his trans-continental trek, in August 1860, with Robert O'Hara Burke. Becker was interested in all things scientific and had encouraged Brown in his thinking. The expedition had departed, however, without anything being done to provide the party with a balloon. Once again Brown had to look elsewhere for his opportunity.

Christopher Pond and Felix Spiers were the proprietors of the popular *Café de Paris* in Bourke Street, and they had an interest in many other entrepreneurial undertakings. In January 1861 their attention was fixed on the task of bringing out the first visiting English cricket team to the Australian colonies, and Coppin was also in the running to sponsor the visit.[7] Brown saw the situation as an opportunity to gain support for yet another attempt to get aloft, while at the same time demonstrating to Coppin how wrong he had been earlier not to have seized the chance of securing his services. His idea was to approach Spiers and Pond with the same idea he had floated the previous year, of arranging a balloon ascent in Tasmania. This time the purpose of the exercise would be to stage an event that would help promote and raise funds for the projected cricket tour. Pond was quite taken by the idea and arranged to meet Brown.[8]

The arrangements to bring out the English cricketers were not going smoothly. The players were happy to tour provided their fees could be met, but others, aside from Coppin, Spiers and Pond, had now come into the bidding to organise the tour. Pond no longer had time to consider Brown's proposal. He and his partner Spiers dispatched their agent to England to procure the services of the best cricketers. A playing-fee of £150, first-class travel and expenses were the enticements offered. The other challengers were unable, or unwilling, to match this handsome offer and by June Pond and Spiers had concluded the deal, with the Melbourne Cricket Club granting them free use of their ground, while retaining the exclusive use of the pavilion and the public bar.[9] Brown heard nothing more from Pond about the Tasmanian project.

By the middle of the year, 1861, the public's attention had turned to matters other than the impending cricket tour. Robert O'Hara Burke, together with some members of his party, had gone missing in the outback. The Royal Society, organisers of the expedition, now formed a committee to undertake the search and rescue of the missing explorers. Brown promptly wrote to Dr John Macadam, the Secretary of the Royal Society, offering to build a hot-air balloon to assist in the search. It would be used as a *ballon-captif* with the object of providing the rescue party with a lofty lookout from which to scan the countryside. At an elevation of 3,000 feet 'a vast extent of country' could be surveyed.[10] The craft would cost in the region of £50, but Brown undertook to join the search and manage the balloon without any further remuneration. The Exploration Committee, as it was named, received Brown's letter. It was read to the meeting, and passed to the leader of the rescue party for his consideration.[11]

Alfred William Howitt had been selected by the Royal Society to lead the rescue party. An Englishman who had only been in Australia for nine years, he had nonetheless gained a reputation for bush-craft. He referred to himself as being 'naturally a savage', with 'the open air and forest both being necessary to my existence'.[12] When news first broke that the Burke party might be in trouble, Howitt had remarked to a friend, 'I suppose if they are lost they will send for me to go out and look for them'.[13] A man who enjoyed the challenge of a hard slog through the bush, the thought of 'flying' held no particular attraction for him. When he read Brown's letter he scribbled on a corner, 'I do not consider the scheme practicable on such a journey as mine,' he added, however, 'a trial should be given by Mr Brown in the Royal Park'.[14] Nevertheless, Howitt was prepared to test out another novelty. He opted to take some carrier pigeons and by using them send back information of the party's findings 'within 24 hours'.[15]

Brown was devastated by this rejection. While Pond had not followed through on the Tasmanian venture, Brown had every hope that he would be involved in the Burke rescue. His earlier conversations with Becker had even led him to believe that a balloon would be part of the original expedition's equipment, and now it was not even allowed as being of value in the rescue attempt. Unwittingly adding salt to the wound, Edwin continued to send details of the successful ascents being made in England by Coxwell and others. It appeared that another year would pass with Brown still firmly planted on the ground. Then Brown was unexpectedly approached by a new player – 'Professor' J.F. Johnston.

Johnston, a pyrotechnist, had been brought out by Coppin at the same time as Brown, but had very quickly gone his own way. He had travelled to India, where his firework displays had proved a success, entertaining settlers and natives alike. He now asked Brown to specify on what terms he would be prepared to make a balloon for use in India. Brown would be under no pressure, as Johnston did not intend to leave Australia for some time. Having thought about the matter, Brown outlined his terms. He would make the balloon at no cost if Johnston provided workmen and a shed in which to build the craft. There was one further condition. Brown would have the machine completed by December and he must have the sole use of the balloon for as long as it remained in Australia – and Johnston had indicated this could be for as long as six months – and he would take half the profits from all ascents.[16] Johnston readily agreed to this arrangement, but his alacrity left Brown with a niggling suspicion. In an attempt to abate his unease, Brown asked Johnston to enter into a signed agreement regarding the construction and use of the balloon. Again, Johnston readily agreed to do so. Brown produced a written contract, and it was taken away by Johnston for a perusal. But Brown remained uneasy and he imagined himself becoming the victim of the other's subterfuge and sharp dealing. He decided to turn the tables. He would take a short cut in the manufacture of the balloon. To this end he opted to use low-priced muslin in the construction of the envelope. The material, when varnished, would be sufficiently robust to sustain a few inflations, but after that would very likely deteriorate quickly. He would get some flights out of it, but he doubted the balloon would survive long in the heat and dust of India. He did not consider his conduct to be irrational in any way.

By mid-December Brown saw his suspicion being realised. While Johnston provided the premises in which the balloon was to be cut and assembled, to Brown's intense irritation, Johnston took over the supervision of the two men who had been employed to varnish the envelope. He also refused to sign the agreement Brown had drawn up, stating the he could not possibly let him have half the profits. He argued that he could get Green or Dean to make ascents for £2 a go.[17] Angered by this – as he saw it – unconscionable conduct, Brown further sabotaged the undertaking by surreptitiously mutilating the pattern he had made. It would be impossible for Johnston to cut any replacement gores for the balloon. Brown seemed not to be aware just how irrational his conduct was becoming. He announced that once the maiden flight of the balloon was over he would have no further involvement with it; but he insisted that he be the one to make the first flight. Realising that he had tested Brown's patience as much as he dared, Johnston agreed.

Notes

1 Brown to Coppin, 26 September 1860, *Correspondence*: 316.
2 Brown to Edwin Brown, 24 November 1860, *Correspondence*: 331.
3 Thomas Rea – claiming in his handbills to be 'the first aeronaut in her Majesty's Colonies' – announced a balloon ascent at Launceston, vide *Melbourne Daily News* 25 December 1849: 4.
4 *Mercury* 16 October 1860: 3.
5 Brown to Coppin, 8 December 1860, *Correspondence*: 331.
6 'Balloons for Exploration', *The Argus* 4 January 1861: 5. 'E.S.B's proposed system proposed a complex gas compression system to allow the gas to be re-used after each ascent. It was not a practical proposal. On the other hand, Brown's simple Montgolfier could probably have succeeded in getting airborne.
7 Alf Batchelder, 'A Closer Look at the Spiers and Pond Tour of 1861–62', *The Yorker*, Autumn 1999: 2.
8 Brown to Pond, 30 January 1861, *Correspondence*: 335.
9 Batchelder 4.
10 Brown to Macadam, 1 July 1861, *Correspondence*: 337.
11 'The Exploration Party', *The Argus* 5 July 1861: 5.
12 Mary Howitt Walker, *Come wind, come weather* (Melbourne: Melbourne University Press, 1971) 79.
13 Walker 116.
14 Endorsed on the original, State Library of Victoria. *Exploration Committee Papers*, 'Correspondence Received' MF 429. Royal Park was the location from which Burke had set off on 20 August 1860.
15 Walker 122.
16 Brown to MacSweeny, 24 October 1863, *Correspondence*: 401.
17 Brown to Coxwell, 25 March 1863, *Correspondence*: 390.

CHAPTER 24

The All England Eleven

After a wait of close on three years, Brown finally had a balloon at his disposal that he could use on his own terms. Granted, it was a machine that had not been built to last. He quickly sought out Pond and offered his services in connection with the visit of the All England Cricket XI, who were already in Victoria and were scheduled to play against a Victorian XVIII in the new year. Pond was happy to engage Brown at a fee of £40, but the balloon would have to be named the *All England Eleven*. Pond would pay for the decoration of the craft.

Left to his own devices Brown quickly made the arrangements for the ascent. The Melbourne Gas Company was prepared to inflate the balloon gratuitously from the mains it had newly laid up to Fitzroy Square.[1] This offer was made not out of altruism, but rather to stop the rival gas company in Fitzroy gaining any publicity and establishing good-will.[2] It was just a short way across Wellington Parade to the Melbourne Club Cricket Ground where the match was to be played. The balloon was elaborately decorated, with a portrait of Queen Victoria painted on one side of the 45-foot tall envelope and the Arms of Australia on the opposite side, and its name lettered around its 116-foot circumference. Pond and Spiers placed advertisements in the newspapers promoting the event, untruthfully boasting that the balloon was 'the first ever manufactured in the colonies, and announcing that 'Professor Brown, the celebrated aeronaut' would be accompanied by a lady.[3]

The lady who was to make the ascent was to be Mrs Johnston. Unfortunately, she had had a blazing row with Brown during the construction of the balloon, and was not on speaking terms with him. Brown asked his wife to accompany him on the flight, and she agreed. Dean was also engaged as a casual helper, even though he had fallen back on his old ways and was drinking more than was reasonable. Still, there was less likelihood that he would attempt to steal Brown's thunder, whereas that was always a present danger if Green were involved in the undertaking.

The cricket match itself had created great interest. The English team had arrived on the *Great Britain* on Christmas Eve, and had been paraded through Melbourne and feted at Pond and Spiers' *Café de Paris*. A new-tiered stand, designed to hold up to six thousand spectators, had been erected at the cricket ground. The promoters' expectations were met when the crowds filled the grounds on the very first day of the match, and continued to do so on each subsequent day, even though the local Victorian XVIII were being outclassed by the English team.

The ascent was scheduled to take place on the last day of the match. Brown was excited by reports that there were likely to be as many as 25,000 spectators present. He had the balloon brought to the Square in the morning, but held off the inflation as long as he could since it was a very hot day, and he was anxious that his jerry-built balloon would lose a great amount of gas by leakage from the craft's seams and through its cheap porous material. By two o'clock word was relayed from the grounds that the Victorian XVIII, batting to reach a total of 187 runs, was in serious trouble. Brown began the inflation.

For a while it appeared that his worst fears were about to be realised. Despite the best efforts of Mr Hutchinson, the gasworks engineer and his crew, the *All England Eleven* remained flaccid. Brown then discovered that Dean, who once again gave signs of being the worse for wear, had rigged the balloon valve incorrectly; it was wedged open instead of being firmly shut. He quickly re-set the valve, and the balloon began to take on gas and become buoyant. An hour and a half later the balloon was filled to capacity. Johnston had invited himself to the event. He arrived on the scene with Pond and announced that it was time to tow the balloon the short distance across Alexandra Parade to the open ground of the Police Paddock, where the Melbourne Cricket Club had its grounds.

As luck would have it, the balloon was brought to the edge of the cricket ground just as the last wicket fell. The crowd rushed across the grounds and surrounded the craft. Pond had seated himself in the car and received an ovation as he was floated to the centre of the ground.

Surrounded by a heaving crowd, Brown and Dean struggled to keep the balloon under control. The band, which had provided occasional entertainment during the match, now struck up *See the Conquering Hero Comes.*[4] Coppin would have been proud of the theatricality of the occasion. The balloon was manoeuvred to the Governor's box at the front of the Grand Stand.

His Excellency Sir Henry Barkly graciously deigned to inspect the craft, though not before Dean had finally proved his worth by manhandling out of the car an intruder who had drunkenly announced his intention of also making the ascent. Set a drunk to catch a drunk, thought Brown. The balloon, dignitaries and aeronauts having been photographed, and the crowd still milling around, an anxious Pond ordered the balloon aloft. Affecting an airy composure, Brown assisted Eliza into the car and with Dean also on board, gave the order 'hands off'. To his satisfaction, and somewhat to his surprise, the *All England Eleven* reacted beautifully. Without any twisting, lurching or drifting, it rose slowly and steadily. An over-excited Johnston seeing his balloon successfully aloft shouted garbled instructions from the ground, but was ignored by Brown who waved nonchalantly to the crowd.[5]

The trio stayed in the air for just over half an hour and landed uneventfully at the north end of Albert Street, near Studley Park. A crowd had followed the balloon and, mindful of his previous experiences at the hands of Melburnians, Brown feared the worst. Seeing Eliza as a passenger, however, the people were boisterous rather than aggressive. The crowd helped Brown and Dean empty and fold the balloon. Spiers and Pond had sent a van to collect the balloon, which was quickly packed and dispatched.[6] Eliza, Brown and Dean boarded a gig and were escorted back to town by a cheering throng. It was the end of a perfect day.

The euphoria generated by this success was soon dispelled when Brown discovered that Johnston had taken receipt of the £40 fee paid by Pond for the ascent. Brown quickly sought out Johnston and demanded his money. Johnston equivocated, arguing he had not intended that Brown should have the entire fee earned from his use of the balloon. He handed over £10 and impertinently suggested that he undertake more ascents for the same fee. Brown was furious and refused to have anything further to do with Johnston or his balloon. Johnston was by now inured to such outbursts and counted them for naught. With regard to the balloon there was little chance of Brown getting involved with it on his terms since Johnston had cannily asked to have the *All Eleven England* delivered to his residence in Richmond after it was recovered from Albert Street. Deprived of the use of a balloon and without any further sponsorship, Brown's aeronautic aspirations were once again frustrated.

Johnston, on the other hand, had grand plans for the future of the balloon. He believed that there was money to be made by exhibiting it on the goldfields before going to India, He managed to achieve something Brown had thought to be unlikely. He secured bookings to make ascents at Ballarat, Sandhurst and Castlemaine. Not being willing to approach Brown to act as pilot, he engaged the services of the ever-willing inebriate, Dean. Anyway, Dean would make the ascent for £5 or less – there would be no haggling about an exorbitant £10 fee! Johnston was well satisfied with his arrangements. To avoid any possibility of a further quarrel with Brown, he secretly shipped the *All England Eleven* to Ballarat.[7] The first Brown knew of the matter was when his attention was drawn to an advertisement that appeared in *The Ballarat Star* early in March:

> This evening. Copenhagen Grounds, Ballarat. Unrivalled out-door attractions! The Gentry and Public of Ballarat and districts are respectfully informed that the great new Victorian Balloon, ALL ENGLAND ELEVEN which ascended from Melbourne Cricket Ground with a lady and two gentlemen, in the presence of His Excellency the Governor, and 25,000 people, will positively make the first grand ascent ever attempted here under the management of the Veteran Aeronaut Professor J. Dean, who it is expected, will be accompanied by a lady and a gentleman. Parties wishing a seat in the car can apply early at the above grounds. In the evening there will be a Superb and Costly Display of fireworks by Professor PRESCOTT of Cremorne.[8]

It was now Johnston's turn to experience frustration. Bad weather caused the cancellation of the flight, and it was two days before it cleared sufficiently to allow for the balloon to be inflated at the local gasworks. The Ballarat Gas Company was situated at the bottom of Dana Street hill, a very steep incline, but which led directly to the Grounds. With the Ballarat West Fire Brigade Band leading the way the balloon was towed up the hill. Dean, directing operations, was drunk. It was a rather battered balloon that finally arrived at the Ground. The *All England Eleven* had been entangled in telegraph poles and dragged along Ballarat's unmade roads. It was now so badly torn and losing so much gas that it proved impossible to get it airborne. In the grip of an alcoholic-induced hysteria, Dean first dumped the grapple hooks and, with the balloon still giving no indication that it would rise, he then cut away all the superfluous rigging. Finally he detached the car and attempted to get aloft clinging to the netting, all without success. Youngsters were sent amongst the crowd carrying placards advising that the ascent would not, after all, take place. Prescott was able to provide the promised fireworks, and surprisingly this was accepted as 'ample compensation for the disappointment those present experienced at first'.[9] The fact that there was a licensed hotel at each corner of the Copenhagen Grounds may also have helped to mollify the crowd.[10]

Charge and counter charge flew after the event. Johnston claimed the gas supplied was 'too heavy', while T.E. Jones, engineer at the gasworks, countered that the balloon was leaking. Both parties were probably correct in their assessments. Johnston abandoned any idea of remaining on the goldfields and cancelled the scheduled ascents at Sandhurst and Castlemaine. Back in Melbourne, Brown received reports of the contretemps with considerable satisfaction. He was not surprised at the failure of the venture. He knew how unreliable Dean was when on the grog, and earlier enquiries had convinced him that the gas being manufactured at Ballarat was not suitable for ballooning, being made of cannel coal and having poor lifting qualities.[11]

Resilience was one of Johnston's more attractive character traits. Within a month of the disaster at Ballarat he had bounced back. He approached Brown once again with the proposal that they jointly undertake the ballooning venture in India – there was no mention of the £30 that he had duped Brown. Stoutly, Brown declined. 'I have no desire to take the balloon in hand again', he told Johnston, well aware that the craft had only a very limited life expectancy. Then, strangely, he made a most generous offer: 'I will give you information that will ensure your success in future, whosoever may be your aeronaut'.[12] He presented Johnston with four tightly written pages of instructions on how to inflate and pilot a balloon, concluding with the observation, 'If you follow the instructions I have given you, you will have nothing to fear but the elements and malicious persons'.[13] All that Johnston needed now to do, before his departure for India, was to engage an aeronaut. Earlier he had a falling out with Green. Anyway, Johnston considered Green an undesirable risk because of his tendency to seek out the limelight. Dean was the only option, but he would be kept firmly in his place. In August 1862, Johnston, his wife, Dean and a number of assistant pyrotechnicians set sail for India. The *All England Eleven*, now re-named *Victoria*, was stored aboard ship.

Brown had no balloon, and he had refused Coppin's offer to fly in the *Tavistock* – even if it could have been made airworthy. Once again the opportunity for any aeronautical endeavour seemed remote. By now Coppin had closed down the Cremorne Gardens because of the poor attendances, and there was no indication that the venue would open again though there were rumours that Coppin was going to sell the place.[14]

Eliza was once more pregnant. She had supported her husband as best she could up to that point, but she now fondly hoped he would put aside his desire to carve out a niche for himself as an aeronaut and secure a steady income for her and the four children. Then in September at a meeting of the Royal Society of Victoria, which Brown regularly attended as a visitor, the Secretary, Dr John Macadam, introduced Michael Costello and his model of an apparatus for steering balloons. He came before the Society on the recommendation of no less a personage than His Excellency the Governor. The Secretary advised the meeting that, for an outlay of £50, Costello 'would be enabled to construct a balloon which would take up himself and another man and thus afford an opportunity for testing his invention'.[15] Brown was flattered when Costello sought him out after the meeting and asked him to join in the project.[16] The Committee of the Society had told Costello that they would communicate with him concerning funding.

The surge of enthusiasm which Brown initially felt quickly dissipated as he came to realise that what Costello proposed was the use of sails and flapping wing devices – all of which had been unsuccessfully trialled by aeronauts of Brown's acquaintance in England. Bitterly disappointed, he wrote to *The Age* pointing out that Costello 'was mistaken in supposing that he can propel his machine in any direction in the open air'.

Later one more opportunity presented itself for Brown to take to the air. William Prescott, the pyrotechnist who had been involved in the Ballarat debacle with Johnston and Dean, was part owner of the Copenhagen Grounds. He was still convinced that the people of Ballarat would flock to see balloon ascents, particularly if they were attended by firework displays, both on the ground and from the balloon. He had experience of arranging such events at the Cremorne Gardens. He needed a balloon for the project. Would Brown build it? Yes, he would, but at a price. He asked for £100 to construct a modest-sized balloon. This was an extravagant quote, but Brown needed the money; Eliza had just given birth to their second daughter, Miriam.[17] The price was too much for Prescott.

Notes

1 The name *Fitzroy Square* was changed to *Fitzroy Gardens* later in the year (1862) when the Government tidied up the legal status of the reserve by designating it as a 'temporarily reserved' site for a public garden.

2 The *Collingwood, Fitzroy & District Gas Company*, had been set up in August 1860, and was situated just north of Fitzroy Square, at Smith & Reilly Streets (now Alexandra Parade.) John Keating, *The Lambent Flame* (Melbourne: Melbourne University Press, 1974) 79.

3 *The Argus* 4 January 1862, sec. Amusements: 8.

4 'The All-England Eleven against Eighteen of Melbourne and Districts', *The Argus* 6 January 1862: 5.

5 Brown to Coxwell, 24 March 1863, *Correspondence*: 391.

6 *The Age* 7 January 1862.

7 Brown to MacSweeny, 24 October 1863, *Correspondence*: 402.

8 *The Ballarat Star* 5 March 1862: 3.

9 *The Ballarat Star* 8 March 1862: 3.

10 J. Hargraves, *Ballarat Hotels Past and Present*: 33. Unpublished Mss. Australiana Reference, Central Highlands Regional Library, Ballarat.

11 Having heard that Dean was in Ballarat with the balloon, 'I made enquiry about the Ballaarat gas – and found it was made from cannel gas, and so very heavy,

that it was useless attempting to fill the balloon with it.' Brown to MacSweeny, 24 October 1863, *Correspondence*: 402.

12 Brown to Johnston, 25 April 1862, *Correspondence*: 343.

13 Brown to Johnston, Undated April 1862, *Correspondence*: 346.

14 By 1864 the Gardens were no more, with part of the site being used as a lunatic asylum and the remainder being subdivided for factories. Vide: Agnes Paton Bell, *Melbourne: John Batman's Village* (Melbourne: Cassell, 1965) 83.

15 'The Royal Society', *The Argus*, 30 September 1862: 5.

16 Costello to Brown, 1 October 1862, *Correspondence*: 349.

17 6 October 1862, entry 1059. *Registry of Births, Collingwood.* On the form of registration Brown gives his profession as 'Aeronaut'.

CHAPTER 25

The Grand Discovery

The letter was dated 8 January 1863 and postmarked Calcutta. Johnston was reporting on his progress in India. 'I was pushing all the advertisements and publicity in my power', he told Brown, and he boasted of having received the patronage of the Viceroy. The letter also contained the news that 'Dean has turned out one of the most shameful drunken pigs you ever heard tell of. Every week since we landed here he has overdrawn his wages and spent every shilling in drink with all the filthy old soldiers and sailors he can pick up with here in Calcutta', complained Johnston.[1] There was, also news regarding the ascent of the *Victoria* on Boxing Day, 1862. 'The Ascent from Eden Gardens', reported the *Bengal Hurkaru*, 'was accomplished successfully', but:

> We regret that in a pecuniary point of view, the spirited Project has been a very heavy loser. The vicinity of the Gardens was crammed in every direction by Natives, anxious to catch a glimpse of the inflated monster, but in no wise anxious to contribute anything towards the heavy liabilities incurred by the owner. This is peculiarly a failing among the lower order of Natives, but we were glad to find that the superior class mustered inside in large numbers.[2]

The last paragraph brought back memories of the *Australasian*'s maiden voyage at Cremorne Gardens.

For all his complaints about Dean, Johnston had left him in charge of the ascent from the Eden Gardens, but he was far from satisfied with the result. Dean elected to keep the balloon in the air for no more than ten minutes, thought it did reach a considerable height.[3] In spite of Brown's attempt at sabotage, the balloon was flying well, and he felt a perverse pride in that fact. Johnston, however, was beside himself. 'If a more satisfactory ascent is not made Dean will *** the reputation of the balloon in

India, so much for the drunken, filthy hog'.[4] In an attempt to recoup his losses Johnston arranged for another ascent, this time from the Maidan at Calcutta. He took charge of the balloon, with Dean assigned the role of assistant. The *Bengal Hurkaru* again noted the event and observed that the Natives could not 'fail to be gratified in watching the calm self-reliant bearing of the Anglo-Saxon, and which for their own sakes they will do well to imitate'.[5] Dean was sober on the day.[6] From there on in Johnston was able to report that things were 'going ahead like steam power', and the money was rolling in.[7] Brown's situation in Melbourne was in no way as rosy.

The Cremorne Gardens had been opened once again, in a brave effort to win back patrons. But the holiday season was no better than the previous year, and on 7 February the Gardens closed for the last time.[8] Brown greeted the news of Cremorne's demise with mixed emotions. The closure deprived him of a ready-made venue from which to make ascents – and while Coppin had not been ready to meet Brown's demands for what he considered unreasonable recompense, he had always treated him decently. On the other hand, there was now no competition in the aeronautical field – unless one cast Costello in that role, and Brown didn't. But Costello would not go away. Brown had discovered that the man was barely literate and knew next to nothing about ballooning. The Royal Society of Victoria had made the same discovery, and there were no funds forthcoming for Costello to develop his project, but he persisted in his scatterbrained scheme, and wanted Brown to build him a balloon that could ascend to an altitude of six miles (31,680 feet) and lift a weight of seven and three-quarter tons. The previous year, Coxwell and Dr James Glashier had ascended from Wolverhampton. The object of the flight was to conduct barometric experiments and they had reached a height of 30,000 feet, but nearly died through lack of oxygen.[9] Brown knew of the near disaster and warned Costello of the dangers. As a further disincentive, he quoted the staggering price of £300 for the manufacture of the balloon.[10] After that the two men broke off contact.

The *Victoria* was providing stalwart service in India, and Johnston made a quick trip back to Melbourne in September 1863. He told Brown of his plans to take the balloon to China. He reckoned that balloon ascents would be well patronised by the international community at Shanghai, even if the natives were not interested. He had used Brown's little handbook on ballooning to good advantage and was enthused with his success as an aeronaut. He asked Brown to consider joining him when his plans were further advanced. The idea would be for them to move between India and China, with perhaps a trip or two to Java.[11]

Eliza was not at all in favour of the project. This was an entirely different matter from ascents made locally or even as far afield as Sydney. How would the children be provided for? Was her husband going to throw over a steady job for this airy-fairy adventure? And what if there was an accident? She was also conscious of her husband's increasing moodiness, and felt he might be sickening. Undeterred Brown boldly asked Johnston to make an advance so that the family could be provided for in his absence, and on that condition he would undertake the trip. The asking price was too high for Johnston, and the offer lapsed.[12]

Johnston took himself back to India with his new equipment and a firm resolve that he would go it alone. At the very worst he would have the bumbling assistance of Dean. Johnston had left him in hospital in Calcutta, recovering from a kick he received from an irritated equine in whose stable he had sought to sleep off the effects of a hard drinking session.[13]

Costello and Johnston were not the only ones with schemes afoot. For years Brown had been working on a project, details of which he had kept secret from all those about him, even from Eliza. Now he could no longer contain himself, and he revealed to Coxwell: 'I have made a grand discovery – a new sort of balloon requiring no gas or fire – a fact!' He then cautioned, 'I do not intend to let out the secret to any person without reaping some benefit from it. I intend to offer the invention for sale'.[14] The discovery was so secret that he kept all the details in his head, and did not advertise his discovery in the newspapers. Unsurprisingly, there were no buyers.

Another Christmas came. There was no balloon for Brown to use.[15] There was no prospect of a partnership with anyone. He would have nothing to do with Green, nor Green with him. He wrote to Edwin in January 1864:

> Every time I see an account of a fresh ascent being made in England, or elsewhere, it makes me long to be 'up and doing' ... I think I acted very foolishly in coming out here, to a field too limited for remunerative ballooning. I think if I had remained in

> England I should have done well – much better than I do here. Sometimes I think I will make an effort to get back, just for one summer; but the expense of the voyage, and having to provide for my family during my absence I cannot manage. If I were single I would be with you very soon.[16]

The studio of Batchelder and O'Neill, photographers, stood at the east end of Collins Street. In the window at that time was proudly displayed an aerial photograph of the city of Boston, as seen from the balloon *Queen of the Air*, taken by James Black of the firm Black and Batchelder, Boston. Time and again Brown stood in front the window amongst the chattering viewers, and contemplated what could have been. Twelve months later the following item appeared in *The Argus*:

> On Tuesday evening last, Constable Delany, a somewhat active member of the Fitzroy police, was walking down Gertrude Street, nearly opposite the Presbyterian Church, when two cabs drove rapidly past. A person in one shouted out to the driver of the other, which was behind, 'There's a drunken man in the street, jump down and lift him onto the footpath.' At the time Delany thought he recognised the voice as that of a driver named W. Clifford. However, the driver of the last cab, Alfred Smeaton by name, jumped down and lifted the drunken man to the footpath, then climbed into his car, and drove off. Delany, who was still at a little distance, hastened up to the inebriate, accompanied by a man named Curtis. Together they lifted up the tipsy personage, who described himself afterwards as on C.H. Brown, a law clerk, and his first words were, 'Where is my watch?' They looked, and the man's watch was gone, and ribbon by which it had been suspended was broken. It was plain that there had been a robbery, and Delany started to see Clifford, whose voice he had recognised. He found the latter, and from information given, started to the Castle Inn, a neighbouring public house. Arriving here he saw a man examining a watch in the light of one of the windows. It turned out to be Smeaton, the cabdriver who had lifted Brown on to the footpath, and the watch was the stolen one. Smeaton gave a number of explanations as to how he became possessed of the watch; but on being tried at the Fitzroy Police Court yesterday, he was adjudged to be guilty, and sentenced to six months hard labour in gaol.[17]

Nothing more is known of Charles Henry Brown up to 18 January 1870.

Notes

1 Johnston to Brown, 8 January 1863, *Correspondence*: 355.
2 'The Balloon Ascent', *Bengal Harkaru and The India Gazette* 27 December 1862: 3.
3 *Bengal Harkaru*: 3.
4 Johnston to Brown, 8 January 1863, *Correspondence*: 357. Brown was left to imagine the expletive.
5 'Calcutta Ascent', *The Bi-Weekly Bengal Huruaru* 5 March 1863: 74.
6 Johnston to Brown, 8 March 1863, *Correspondence*: 380.
7 Brown to Coxwell, 25 March 1863, *Correspondence*: 388.
8 Alec Bagot, *Coppin the Great* (Melbourne: Melbourne University Press, 1965) 240.
9 Erik Norgaard, *The Book of Balloons* (New York: Crown Publishing Inc., 1971) 64. Brown was aware of the detrimental effects experienced at high altitudes but he mistakenly believed that Coxwell and Glashier suffered the ill effects they did as a result of gas poisoning.
10 Brown to Costello, 29 July 1863, *Correspondence*: 399.
11 Brown to Edwin Brown, 25 January 1864, *Correspondence*: 419.
12 Brown to Johnston, 21 December 1863, *Correspondence*: 338.
13 Brown to MacSweeny, 24 October 1863, *Correspondence*: 418.
14 Brown to Coxwell, 25 March 1863, *Correspondence*: 396.
15 Somewhere along the line the already faulty *Tavistock* had been destroyed.
16 Brown to Edwin Brown, 25 January 1864, *Correspondence*: 418.
17 *The Argus* 23 December 1864: 4.

In Melbourne

'The House That Jack Built' was having a great season. We were in our fourth week and there was no sign of the audiences dropping off. Mister Coppin was very pleased with the houses we were getting. I didn't see much of him, though. He had all those meetings and such to go to. Mister Henning was the one I saw most, but that's because he painted all the scenery – as well as being the manager. He did some lovely stuff. 'Visual delights' the papers called them. The Theatre Royal is a fine venue. I reckon it's as good as anything we've got in London. But let me get on with my story.

I'd gone to Mickey John Barry's hotel in Rathdowne Street. The Australian he'd called it. Leastways he resisted the temptation to name it The Shamrock or The Brian Boru. He's a good sort for an Irishman. The hotel was just around the corner from my lodgings, and about ten minutes away from the theatre. It was a grand place to come for a spot of lubrication for the vocal cords. I'd a fair bit of singing to do in the pantomime.

I'd got myself snug in the booth, and I gave Mick a wave. He knew what to bring me. I'd have two nobblers of brandy, and then I'd take myself off and have a rest before the performance started at seven o'clock. Mick brought the brandy.

'Thanks, Mick.' I saluted him with the glass. 'Wait a moment.' Mick slid in beside me in the booth. 'Tell me,' I asked, 'who is that fellow over there at the end of the counter?' Mick looked in the same direction as me.

'He's a regular. Lives just up the road.'

'Yes, I've seen him here before.' I'd made a habit of this midday tipple since I'd come to Melbourne to do the pantomime. 'He looks familiar.'

'Familiar?'

'Yes. He reminds me of someone. Can't for the moment think who.'

'Drink your drink,' said Mick, standing up and moving back to the bar. 'It will come to you,' he said over his shoulder. He went behind the counter and talked to the man. They looked back towards me. I got up and went over.

'Can I shout you a nobbler?' I asked, leaning on the counter beside him.

'Do I know you?' he asked me.

'I don't know, but I think I know you.'

'Well, if you're buying the man a drink,' said Mick, 'get a move on. He's done his tick here for the day, haven't you, Brown?'

I bought the man a brandy. 'You're Mister Brown then?'

'Yes, Charles Henry Brown.' He put down the glass and rubbed one eye and then the other.

I thought to myself. Brown … brown … brown … balloon. That's it! Yes, and the accent was there. 'You're from Yorkshire,' I told him.

'Many years ago,' he said, and sipped his drink.

'You're a …' I struggled for the word, and gave up. 'You fly balloons.'

He straightened up and turned towards me. There was a glint in his watery eyes. 'Do I know you? You do seem to know me.' He was very well spoken. Well, he was a little slurred, but well spoken nonetheless.

'I wouldn't expect you'll remember,' I told him, 'but I met you just before you came to Australia. You were in London.'

He turned back and leant on the counter. The light was gone from his eyes. 'Well, this is a long way from London, and a long time from when I left old England.'

'Come and sit down,' I invited, and pointed towards the booth. He hesitated. 'I'd be pleased to buy you another drink,' I offered. He came with me slowly, and we sat. I waved to Mick and he brought over the bottle of brandy. 'Here, you keep a tally,' he said to me. 'You're the one with the cash.' He patted me on the shoulder as he says that, and then had a good hard look at Mister Brown. 'Are you all right, then?'

'Perfectly so,' said Brown, leaning back against the wall of the booth. Mick left us. 'Kind of him to show concern for my welfare,' said Brown. I don't know if that was a joke or if he was being serious. He looked hard at me. 'I think I may remember you.'

I helped him. 'I was making my way in vaudeville. I was engaged to sing at the old Winchester Music Hall. You remember?'

'Perhaps,' said Brown.

'Well, I'm out here with the Duvalli sisters. We're doing a pantomime season at the Royal and then moving on to Sydney.'

'Interesting,' said Brown, but clearly he didn't mean it. His glass was empty, so I poured him another drink. He looked at me. 'Your endeavours in this part of the world have patently proved more rewarding than have mine.'

'What do you mean?' I asked him. He paused for a moment.

'I have had but little success as an aeronaut. Oh, there was an initial burst of interest in aerostation. But then things went awry.'

'Awry?' I asked.

'Accidents,' he said. 'And worse, betrayal by partners, and abandonment by friends. You may not have come against it yet, but I would warn you – double-dealing is a way of life in this colony. No! In this country!'

'I have not found it so, so far,' I answered.

'Oh, but a little more time here, then, and you will.' He had become quite agitated. 'Even I have fallen into that misbegotten way.'

I was quite uncomfortable with the turn the conversation had taken. I'd simply wanted to renew acquaintance with a fellow Yorkshireman.

'I could have made a success of the business,' he continued, and he helped himself to another nobbler. 'I could have saved O'Hara Burke.' I didn't know what he was talking about. 'I would have been a hero, without doubt. But Howitt would not have me on the trip.'

'I have an early rehearsal,' I lied, 'I must be off. Keep the bottle. I will settle with Mick as I leave.' Brown grabbed my sleeve.

'I have an invention,' he whispered. 'I have devised a means of flying which needs neither gas nor fire.'

'Splendid!' I whispered back, and freed my arm.

'If I had but funds.' It sounded as if he were pleading. 'I have five children.'

I quickly crossed to the counter, gave Mister Barry five shillings for the brandy and left. The next day I seriously thought about giving the hotel a miss, but then I thought, why put myself out because of a cranky bugger. So I went for my usual nobbler, but I never saw Brown again. Anyway, the run finished on 20 January and we went to Sydney.[1]

* * *

At 9 o'clock on the evening of 18 January 1870 Charles Henry Brown stood on the banks of the Yarra River under the Cremorne Railway Bridge. He had made a decision. Taking off his hat he balanced it atop a fence. He then removed his pipe and tobacco pouch from one pocket and carefully placed them in the hat; from the other pocket he took a pencil and a battered quill. These he also put into the hat. As he shrugged off his coat, a thought crossed his mind. He recalled how, all those years ago, a bitter debate had raged in the newspapers about the propriety – or impropriety – of bathers displaying themselves in their undergarments; and how that blasted debate had kept some of his letters about ballooning out of print. He smiled bitterly to himself and pulling back on his coat, he stepped into the Yarra.

Note

1 The background details for this vignette were obtained from *The Argus*, 18 January 1870, and Margaret Williams' *Australia on the Popular Stage: 1829–1929* (Melbourne: Oxford University Press, 1983).

Bibliography

Books

Altick, Richard. *The Shows of London*. Harvard: Harvard University Press, 1978.

Anderson, John. *A History of Aerodynamics*. Cambridge: Cambridge University Press, 1997.

Bacon, Gertrude. *Memories of Land and Sky*. London: Methuen & Co. Ltd., 1928.

Bacon, John M. *Dominion of the Air*. London: Long, 1911.

Bagot, Alec. *Coppin the Great*. Clayton: Melbourne University Press, 1965.

Ballantyne, Robert. *Up in The Clouds or Balloon Voyages*. London: J. Nisbet, 1892.

Barber, John. *The Overland Guide Book*. London: W.H. Allen, 1850.

Bell, Agnes Paton. *Melbourne – John Batman's Village*. Melbourne: Cassell, 1965.

Bennett, D. *Encyclopaedia of Dublin*. Dublin: Gill & Macmillan, 1991.

Bevan, I. *The Story of the Theatre Royal*. Melbourne: Hawthorn Press, 1993.

Briggs, Asa. *Victorian Cities*. Harmondsworth: Penguin Books, 1980.

Brogden, Stanley. *History of Australian Aviation*. Melbourne: Hawthorn Press, 1960.

Broughton, T, & Parnell, N. *Flypast*. Canberra: AGPS Press, 1988.

Byrne, Liam. *History of Aviation in Ireland*. Tallagh: Blackwater Press, 1980.

Cannon, Michael. *Life in the Cities – Australia in the Victorian Age*. Melbourne: Thomas Nelson Ltd., 1975.

Carroll, Brian. *Richmond and East Melbourne Sketchbook*. Melbourne: Rigby, 1976.

——. *Australian Aviators*. Melbourne: Cassell, 1980.

Catchpole, Brian. *Balloons and Buccaneers – Yorkshire's Role in Aviation since 1785*. York: Maxiprint, 1995.

Chandler, Charles. *Balloon and Airship Gases*. New York: The Ronald Press Co., 1926.

Copley, G. *Australians in the Air*. Adelaide: Rigby Ltd., 1976.

Coxwell, Henry. *My Life and Balloon Experience, Vol. II*. London: W.H. Allen & Co., 1889.

Cutler, Carl. *Greyhounds of the Sea*. New York: Private publication, 1930.

Dullfus, C. *Up in a Balloon*. New York: Prentice Hall, 1962.

Elliot, Christopher. *Aeronauts and Aviators*. Laverham: Dalton, 1871.

Haldane, Robert. *The People's Force*. Melbourne: Melbourne University Press, 1995.

Hall, H, & Cripps, A. *The Romance of the Sydney Stage*. Sydney: Currency Press, 1996.

Hardingham, George. *Practical Aeronautics*. London: E & F.N. Spon, 1871.
Hassam, Andrew. *Sailing to Australia – Shipboard Diaries by Nineteenth-Century British Emigrants*. Melbourne: Melbourne University Press, 1995.
Heaton, J.H. *Australian Dictionary of Dates and Men of the Time*. Melbourne: Geo. Robertson, 1879.
Hibbert, Christopher. *Social History in Victorian Britain*. Melbourne: Angus & Robertson, 1975.
Hodgson, J.E. *The History of Aerostation in Great Britain*. London: Oxford University Press, 1924.
Hubbard, Thomas. *The Aeronautical Classics*. London: Aeronautical Society of Great Britain, 1911.
Joy, William. *The Aviators*. Sydney: Golden Press, 1965.
Keating, John. *The Lambent Flame*. Melbourne: Melbourne University Press, 1974.
Kelly, William. *Life in Victoria, Vol. II*. Kilmore: Lowden Publishing, 1977.
Lawrence, William J. *The Life of Gustavus Vaughan Brooke*. Belfast: W & G Baird, 1892.
Mackworth-Praed, Ben. *Aviation-The Pioneer Years*. London: Studio Editions, 1990
Maxwell, C. *Dublin Under the Georges*. Dublin: Gill & Macmillan, 1979.
Norgaard, Erik. *The Book of Balloons*. New York: Crown Publishing Inc., 1971.
O'Donnell, A. *The Rathmines Township – A Chronology and Guide to Sources of Information*. Dublin: Dublin Public Libraries, 1998.
Parnell, Neville, & Broughton, Trevor. *Flypast*. Canberra: AGPS, 1988.
Perry, John. *Up, up and away: An Account of Ballooning in and around Bristol. 1784–1999*. Bristol: Bristol Historical Society, 1999.
Proudley, Ray. *Gas in Ballarat*. Melbourne: Gas & Fuel Corporation of Victoria, 1985.
——. *Circle of Influence*. Melbourne: Hargreen Publishing, 1987.
Rickard, John. *A Family Romance – The Deakins at Home*. Clayton: Melbourne University Press, 1996.
Robinson, Brian. *Aviation in Manchester*. Manchester: Royal Aeronautical Society, 1977.
Rodgers, Helene. *Early Ballooning in Australia*. Melbourne: Set and Forget Press, 1989.
Rolt, Lionel. *The Aeronauts: a History of Ballooning 1783–1903*. London: Longman, 1966.
Saundby, Robert. *Early Aviation: Man Conquers the Air*. London: Macdonald & Co., 1971.
Sproule, Anna. *Port Out Starboard Home*. Poole: Blandford Press, 1978.
Stables, Gordon. *Leaves from the Log of a Gentleman Gipsy*. London: Jarrod & Sons, 1891.
Stephens, L, ed. *Dictionary of National Biography*. Vol. XX. London: Smith, Elder & Co., 1889.

Stocqueler, J.H. *A Guidebook to Messers. Grieve & Telbins Diorama of the Ocean Mail to India & Australia Etc.* London: n.p., 1853.

Walker, Mary Howitt. *Come Wind, Come Weather*. Melbourne: University of Melbourne Press, 1971.

Waller, Charles. *Magical Nights at the Theatre*. Melbourne: Gerard Taylor Productions, 1980.

Walton, Derek. *Northumberland Aviation Diary: Aviation Incidents from 1790 to 1999*. Norav: Seahouses, 1999.

White, Milliam. *White's 1853 Leeds & the Clothing Districts of Yorkshire*. New York: A.M. Kelly, 1853.

Williams, Margaret. *Australia on the Popular Stage 1829–1929*. Melbourne: Oxford University Press, 1983.

Wise, John. *A System of Aeronautics, Comprehending Its Earliest Investigations, and Modern Practices and Art*. Philadelphia: J.A. Steel, 1850.

Woodham-Smith, C. *Queen Victoria – Her Life and Times, Vol. I*. London: Hamish Hamilton 1972.

Articles

Batchelder, Alf. 'A Closer Look at the Spiers and Pond Tour of 1861–62.' *The Yorker* [Melbourne] Autumn 1999: 2.

Campbell, J.P. 'Two Memorial Dublin Houses.' *Dublin Historical Record* [Dublin] II (1939): 141.

Coxwell, Henry. 'Exploration of Australia by Balloon.' *Mechanics' Magazine* [London] 28 January 1859: 79.

Dixon, F.E. 'Ballooning in Dublin.' *Dublin Historical Record* [Dublin] XIV (1955): 10.

Horne, R.H. 'Ballooning.' *Household Words* [London] 1851: 83.

Milbourne, Christopher. 'Anderson in America.' *MUM* [New York] *44 (*1954).

O'Neill-FitzSimons, Terence. 'A Balloon on the Ballarat.' *Victorian Historical Journal* [Melbourne] 71 No. 1 (March 2000): 27.

Manuscript Sources

Brown, Charles Henry. *Correspondence*. MSS 9515 Box 2493/1, La Trobe Australian Manuscript Collection, State Library of Victoria, Melbourne.
Coppin, George. *The Coppin Papers*. MSS 8827 Box 12/18, La Trobe Australian Manuscript Collection, State Library of Victoria, Melbourne.
Dodge, Frederick J. *Swiftsure Plymouth to Melbourne, 17 October 1857–22 December 1857*. MSS 11530 Box 1768/1, La Trobe Australian Manuscript Collection, State Library of Victoria, Melbourne.
Griffith, Charles J. *Diary Written on Board the Simla*. MSS 10875, La Trobe Australian Manuscript Collection, State Library of Victoria, Melbourne.
Mattison, Alfred. *Chronicles of Leeds*. MSS 942.75M435, Leeds Public Library.
Municipal Council Ballarat, General Correspondence, 1857–58, VPRO 2500.
Municipal Council Ballarat East, Minute Book, 12–15 January 1858.
Police Department Records, Inquests – Richmond, VPSR, 1870.
Shipping Lists, Immigration Returns, VPSR, 1857/8.

Newspapers

Age [*The*], (Melbourne), 1858–1870
Argus [*The*], (Melbourne), 1856–1870
Ballarat Star [*The*], 1857–1862
Ballarat Times, 1856–1862
Bell's Life in Sydney & Sporting Reviewer, 1859–1860
Bell's Life in Victoria, 1858–1859
Bendigo Advertiser, 1858
Bengal Hurkaru & The India Gazette, 1863
Bi-Weekly Bengal Harkaru, 1863
Bradford Post, 1857–1858
Cork Examiner, 1849–1852
Dublin Evening Post, 1852–58
Era (London), 1859
Evening Packet & Correspondent (Dublin), 1850

Freeman Journal (Dublin) 1849–1850
Herald (Melbourne), 1857–1860
Illustrated Journal of Australasia, 1858
Illustrated London News, 1858–61
Illustrated Melbourne Post, 1862
Leeds Intelligencer, 1856–1858
Limerick and Clare Examiner, 1851
Mechanics Magazine (London), 1859–1860
Melbourne Daily News, 1849
Melbourne Evening Mail, 1858
Melbourne Punch, 1858
Mercury (Hobart Town), 1860
Miner & Weekly Star [*The*], (Ballarat), 1856–58
Saunder's News Letter (Dublin), 1848
Sydney Morning Herald, 1858–1870

Newspaper quotes are courtesy of The British Library, National Library of Australia, and the State Library of Victoria.

Index

Aerostatic Magazine 86, 110, 199
Aerostatic Remembrancer, The 112
Albion 28
All England Eleven 211–15
Anderson, John 119, 128, 149–50, 174, 197–8
Australasian 128, 149, 151, 153, 165, 167, 172, 175–6, 185, 187–8, 197–8

Ballaarat West 154–5
Ballarat 150, 153, 163–4, 215
Ballarat Gas Company 214
Barkly, Sir Henry 154, 213
Becker, Dr Ludwig 204–6
Bell, Hugh 77–9, 81
Bellin, Mr 185, 203–4
Bonthellier, Edouard 95
Bowser, Robert 97
Bradford 16, 27, 51, 59–60, 73–4, 77, 102, 104
Brooke, Gustavus 117, 149, 155, 163, 173–5, 193
Brown, Benjamin 16–17, 29, 94, 110
Brown, Charles Henry: ascents 87, 168, 173, 213;children 35, 58, 95, 142, 153–4, 182, 199, 217; crippled arm 58; dissolution of partnership 195; drowning 3; employment 4, 35, 42, 74, 83, 171, 193; failed ascents 63, 76; grand discovery 221; inquest 9–12; insolvency 199; manuscript 40; marriage 35; parents 16–17; partnerships 34, 50, 75, 116, 172, 180; printer 110; robbed 222
Brown, Edwin 16, 82, 110, 198, 207, 221
Brown, Eliza 5, 9, 35, 58, 110, 127, 147, 171, 176, 182, 199, 213, 216, 221
Burke, Robert O'Hara 206–7

Candler, Samuel Curtis 1–3, 9, 11
Carrick, Susan 8
Celeste 76
Chambers, Stephen 83, 94, 96–102, 116
Charlie Napier Theatre 155
China 220–1
Clapham, Thomas 83, 94, 99, 110
Colossal Figure 103
Copenhagen Ground 214–15
Coppin, George 115, 117, 148, 168, 171–2, 183, 193, 199, 203–5
Coronation Balloon 19
Costello, Michael 216, 220–1
Courtis, Henry 155–6
Coxwell, Henry 17, 27, 57, 85, 87, 112, 117, 151
Cremorne Gardens, London 84, 103, 115, 151
Cremorne Gardens, Melbourne 116, 118, 149, 164, 165, 168–76, 216, 220

Dean, Joseph 118–9, 148, 151, 165, 167, 171–3, 175, 200, 211, 213–7, 220
Delcourt, Dupuis 74–5
Dennison, Sir William 173

Dublin 28–9, 40, 42, 47, 49–51
Dunne, Joseph 158

Edwards, Harry 173–4, 189
Erin Go Bragh 28, 30–1
Exploration Committee, The 207

Field, Jospeh 194
Franklin, Lady 46, 49, 52
Franklin, Sir John 39, 41, 52

Gale, George Bercher 19, 40, 54
Glashier, Dr James 220
Gore, Henry 109
Great Northern 19–20
Green, Charles 16, 28, 94, 112, 154
Green, William 19, 94, 183, 195

Hampton, John 15, 28, 94
Hayes, Richard 86–7
Horne, Richard 168–9
Howitt, Alfred William 207

India 208, 214, 216, 219–21

Johnston, J.F. 207–9, 212–16, 219–21
Jones T.E. 215

Leeds 15–6, 20, 27, 29–30, 35, 42, 49, 51, 84, 96
Leeds Royal Gardens 87, 101
Lunardi, Vincent 16

M'Lerie, Captain 188
Macadam, Dr John 207, 216
MacSweeny, John 17, 57, 75, 82, 98
Maigre, Pierre 154
Melbourne Gas Company 151, 211
Monster Balloon 60, 62, 74
Montgolfier 27, 30, 35, 60, 73, 75, 86, 205

Napoleon 105–6, 116
Neeson, Titus 7, 10
New Artic Surveying Balloon 52, 58, 61, 74
Nottingham Arboretum 103, 105
Nutt, Edward 5

Ormond, Alfred 32, 34, 81
Osburn, William 15
O'Shanassy, Sir John 7

Polack, Abraham 186, 189
Pond, Christopher 206, 211, 213
Poole, Charles 186, 190
Portobello Gardens 28–9
Prescott, William 214, 217

Queen of the Air 220

Rainbow, The 97
Rand, William 60, 73, 77
Rotunda Gardens 28, 52
Royal Gardens 83, 85, 93, 96
Royal Pavilion Gardens 93
Russum, John 62
Russum, William 17, 19–20, 62

Shoberl, William 40–1
Spiers, Felix 206, 211, 213
Standish, Frederick 7, 11–12
Stewart, William 104–7, 111
Surrey Zoological Gardens 93
Sylph 87, 93, 102, 105, 115

Tasmania 204, 206
Tavistock 115, 119, 128,150, 176, 182, 185, 190, 194, 199, 207, 216
Theatre Royal, Ballaarat 163
Theatre Royal, Melbourne 117, 148, 175, 194
Trapeze 86, 94, 112–13

Vauxhall Gardens 186
Victoria, Queen 32, 212
Victory 84, 99, 102
Wadman, William 97
Wardell, Alfred William 186–7, 189
Wilson, Sir Thomas 111
Wise, John 54, 58, 75

www.ingramcontent.com/pod-product-compliance
Ingram Content Group UK Ltd.
Pitfield, Milton Keynes, MK11 3LW, UK
UKHW020213290926
15470UKWH00001B/165